燒味 滷水 小吃

Juicy Barbecue Meat
Enjoy!

序

烹製燒味食品在一般人心中視為難度之製作，由於燒味包含燒、臘、滷、燻等烹調方法，看似十分複雜，而且家中因設備及工具未能配合製作，因此自家製作燒味似乎難度極高，故讀者因而打消自家製作的念頭，只在酒樓食肆及燒臘店才品嘗其美味。

在一個偶然機會下重遇相識二十多年的陳永瀚師傅，一見如故，甚至多番合作機會。陳師傅是一個對工作認真及熱誠的人，而對燒味製作更有獨特的風格及心得，因此很希望與陳師傅合作，將繁複的燒味方法簡單化，讓讀者在家一展身手，做出美味的燒味食品給家人及朋友品嘗，分享歡樂！

此食譜經過討論、考量及製作，終於誕生了！書內除了包括傳統燒味菜式之外，還有從傳統中帶出新意的菜式，令人耳目一新！希望讀者對燒味製作有另一番新看法。

在此，多謝圓方出版社全人及 Rachel 老師在製作時的一切協助及寶貴意見！

即時睇片

序

真的很榮幸！

今次有機會與廖教賢師傅攜手製作這本燒味食譜，多年來的教學心得，令我們更了解初學者的學習困難與顧慮，所以在書內作出相對性的分析說明，並將餐館繁複的烹製過程，轉變為簡易的家庭式做法。

這次與廖師傅合作的特色，是將我們兩人不同的烹飪風格，互相融合、再一起創作。書內設定五個類別：家常菜式、宴客菜式、節日菜式、Party 小食及海鮮菜式。當中的食譜有懷舊的、也有創新的；有適合小朋友參與的親子製作；有多元化的燒味食品，以配合各類場合宴客的需要。

希望這本書，讓你心愛的家人和朋友感受到，家庭式燒味帶來的烹飪樂趣與驚喜！

即時睇片

目錄
Contents

Home-style recipes

Banquet recipes

Seafood recipes

Festive favourites

燒味・小食篇

Snack and appetizers

＊ 除特別標註外，
食譜以 6 人份量為準。

靈活運用燒味工具
Barbecue and marinating tools

鵝尾針

醃製禽類食材時，需要將醃料放入內腔烹製，可用鵝尾針穿縫尾部缺口，以免醃料溢出，而且又可作為鬆針之用途。

Metal skewer

Also known as "goose tail needle" in Chinese, it seals the seam at the tail of a goose after marinade is spread on the inside of the poultry. It prevents the marinade from leaking. You can also use it to prick holes on the skin.

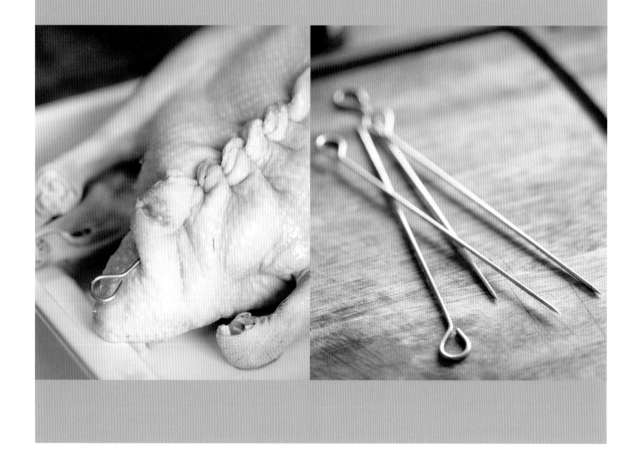

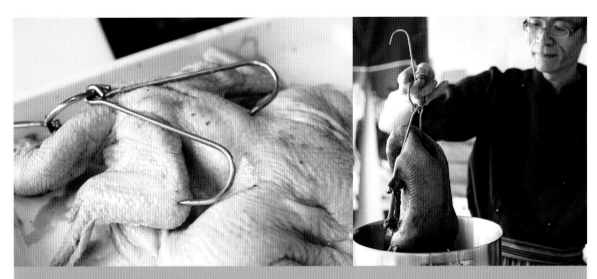

短鴨勾

　　此勾可用作晾掛大型禽類食材，如雞、鵝、鴨及乳豬，掛起後風乾，或勾起已滷製的食品。由於叉燒用鋼針穿連後，需用此環勾掛起放入烤爐，因此又稱為叉燒環。

Short duck hook

　　This is a ring like hook on which larger birds (such as chicken, duck and goose) or big pieces of meat (such as suckling pig) are hung and air dried. The hook can hang the marinated meat. As this hook is also used to secure barbecue pork on metal skewers, this is also known as Char Siu ring.

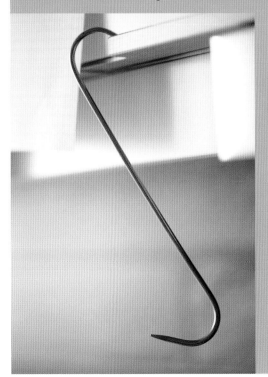

不銹鋼雞勾

　　當禽類食材掃抹上皮水後，作為掛起晾乾之用，一般用於體形較輕禽類的頸部或翼底位置，如鵪鶉、乳鴿及春雞。

Stainless steel chicken hook

　　After applying basting sauce on the skin of poultry, hang it on the hook to let it air dry. Generally speaking, it is used on smaller birds on the neck or under the wings, such as quail, squab or spring chicken.

圓針插

一般用於皮厚或肉質堅韌的食材上，作為鬆針之用途，如五花腩、牛腱或豬手等。

Skin pricker

This looks like a meat tenderize with many needles, but it is much sharper. It is used to prick holes on chewy and rubbery food, such as pork belly, beef shin or pork trotter.

小膠掃

作為上皮、掃上蜜糖及醬料之用途，可均勻塗抹，如烹調脆皮腩仔、紅燒乳鴿、脆皮鰻魚或南乳吊燒雞等。

Rubber spatula

This is used to apply basting sauce, honey glaze and sauce evenly on the food, such as roast pork belly, deep fried squabs, crispy honey-glazed eel or roast chicken in red tarocurd sauce.

常備材料，烹調好輕鬆
Condiment staples that saves time and effort

Basting sauce A

由於糖分的含量比較少，所以適合用於偏高溫油炸的菜式，如腐乳紅燒乳鴿、南乳吊燒雞、香草汁燒乳豬及脆皮鰻魚等。貯存於雪櫃保鮮，可保持最佳效果。

This sauce contains less sugar and is perfect for food that is deep-fried over high heat, because it is less likely to burn. Use it on fried squab with white fermented beancurd sauce, roast chicken in red tarocurd sauce, rosemary roast suckling pig and crispy honey-glazed eel. Just make a whole batch and store in the fridge. It lasts well without affecting the taste.

材料 Ingredients

清水	75 克
麥芽糖	20 克
白醋	80 克
大紅浙醋	5 克
雙蒸酒	45 克
玫瑰露酒	5 克
75 g	water
20 g	maltose
80 g	white vinegar
5 g	red vinegar
45 g	double-distilled rice wine
5 g	Chinese rose wine

做法 Method

清水與麥芽糖混和，座於熱水內至麥芽糖完全溶化，加入白醋及大紅浙醋，待冷卻後，加入雙蒸酒及玫瑰露酒拌勻即成。

Mix water with maltose. Put it over a pot of simmering water and stir until maltose dissolves. Add white vinegar and red vinegar. Wait till it cools. Stir in rice wine and Chinese rose wine.

風乾小提示 Air-drying tip

很多菜式塗抹上皮水後，需要進行風乾的步驟，一般使用烘爐烘乾、電風扇或電風筒吹乾。由於天氣因素直接影響風乾的效果，特別是潮濕的環境，故要適當選擇風乾的方法，其中以電風筒吹乾的效果較好。此外，食品表皮必須吹乾至乾爽及不黏手為佳。

Most barbecue recipes demand the ingredient to be air-dried after the basting sauce is brushed on. Generally speaking, you can do with an oven, an electric fan or a hair dryer. Yet, please note that the weather condition, especially the humidity of the atmosphere, directly affects air-drying speed. You have to pick the right way to dry your ingredients according to the season and temperature. I personally prefer a hair dryer for the best result. Besides, make sure the outer surface of your ingredient is completely dry to the touch before proceeding to the next step. Just test it with your finger. It should feel dry without a hint of stickiness.

Basting sauce B

由於糖分的含量比較多，適合中至偏高溫油炸的菜式，如京式片皮鴨及脆皮琵琶鵪鶉，貯存於雪櫃保鮮，可保持最佳效果。

As this is a sugary sauce, it's mostly used on food that is deep-fried over medium-to-high heat, such as Peking duck and crispy "Pipa" quail. You can make a whole batch and store in the fridge. It lasts well without affecting the taste.

材料 Ingredients

白醋	120 克
麥芽糖	50 克
大紅浙醋	5 克
雙蒸酒	5 克
120 g	white vinegar
50 g	maltose
5 g	red vinegar
5 g	double-distilled rice wine

做法 Method

白醋、大紅浙醋及麥芽糖混和，座於熱水內至麥芽糖完全溶化，取出冷卻後，加入雙蒸酒拌勻即成。

Mix white vinegar, red vinegar and maltose together. Heat it up over a pot of simmering water while stirring until the maltose dissolves. Leave it to cool. Stir in rice wine.

Unforgettable sauce

回味醬

由多種醬料及調味料調配而成，香味濃郁，可作為醃料、調味料或蘸汁食用。

A delectable blend of sauces and condiments, this sauce has potent aroma and robust flavours. Use it as a marinade, a seasoning or a dip.

材料 Ingredients

生油	2 湯匙
海鮮醬	60 克
磨豉醬	60 克
蠔油	20 克
生抽	60 克
花雕酒	12 克
玫瑰露酒	12 克
南乳	10 克
砂糖	140 克
芝麻醬	40 克
2 tbsp	cooking oil
60 g	Hoi Sin sauce
60 g	ground bean sauce
20 g	oyster sauce
60 g	light soy sauce
12 g	Shaoxing wine
12 g	Chinese rose wine
10 g	red tarocurd
140 g	sugar
40 g	sesame paste

做法 Method

1. 將生油及芝麻醬各自用碗盛起；其他材料混和備用。
2. 用生油起鑊，加入已混和的調味料，用慢火煮滾，最後慢慢倒入芝麻醬拌勻，煮至翻滾即成。

1. Set aside the cooking oil and the sesame paste separately. Mix the rest of the ingredients.
2. Heat a wok and add cooking oil. Pour in the sauce mix from step 1. Bring to the boil over low heat. Stir in the sesame paste slowly at last. Bring to the boil.

薑茸 Ginger and spring onion dip

燒味菜式中常見之蘸料，薑味香濃，若貯存於雪櫃，不宜放入葱混和，以免變壞，待食用時才加入葱調味。

The most popular and common dip for barbecue and marinated chicken, this condiment is spicy without being piquant. If you intend to make a big batch and store in the fridge, do not add the spring onion because it tends to turn stale quickly. Just put in freshly chopped spring onion before you serve.

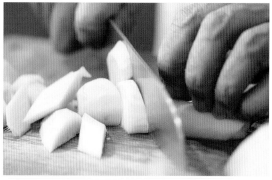

材料 Ingredients

薑茸	4 湯匙
葱茸	4 湯匙
生油	6 湯匙
幼鹽	半茶匙
雞粉	半茶匙
4 tbsp	grated ginger
4 tbsp	finely chopped spring onion
6 tbsp	cooking oil
1/2 tsp	table salt
1/2 tsp	chicken bouillon powder

做法 Method

1. 薑用刀切或攪拌機攪碎成薑茸，放入煲魚湯袋內，壓去薑汁（薑汁貯存留作烹調其他菜式之用）。

2. 薑渣盛起，加入幼鹽、雞粉及葱茸拌勻，倒入燒熱的生油拌勻調味，即可食用。

1. Chop the ginger with a knife or get it done with a food processor. Put it into a muslin bag. Squeeze the extract the ginger juice (you can save it for other dishes).

2. Put the squeezed ginger into a dish. Add salt, chicken bouillon powder and spring onion. Mix well. Heat cooking oil until smoking hot. Pour it over the ginger mixture. Serve.

Steamed shiitake mushrooms

熟冬菇

已處理的熟冬菇用途甚廣，可作菇片、菇絲及菇粒等配料，每次用筷子取出，以免受污染，隨時取用。貯存時不要加入太多冬菇蒸汁，或不要也可，冷藏於雪櫃保存。

Shiitake mushrooms are widely used in Cantonese cuisine, sliced, shredded or diced. Yet, it takes time to re-hydrate and needs some work to be seasoned. Thus, you can make a whole batch and store it in the fridge. Just use a pair of clean chopsticks to take out what you need for one recipe so that the remaining mushrooms won't be contaminated. Make sure you don't soak the mushrooms in too much sauce. If you want, you can actually drain them well so that they are relatively dry to last longer in the fridge.

材料 Ingredients

乾冬菇	300 克
鹽	1 茶匙
生油	2 茶匙
花雕酒	1 茶匙
薑	3 片
葱	1 條
300 g	dried shiitake mushrooms
1 tsp	salt
2 tsp	cooking oil
1 tsp	Shaoxing wine
3 slices	ginger
1 sprig	spring onion

做法 Method

1. 乾冬菇用清水浸過面，浸約 3 小時至軟身，剪去硬蒂，用筲箕盛起，倒入生粉 2 湯匙洗擦表面污漬，洗淨，擠乾水分。

2. 冬菇放入窩內，用清水蓋過面，放入其他材料隔水蒸約 1 小時，待涼，放入密實盒內貯存，可保存 1 至 2 星期。

1. Soak the mushrooms in water (enough to cover) for 3 hours until soft. Cut off the stems. Put them into a strainer. Add 2 tbsp of caltrop starch and rub well to remove the dirt. Rinse and squeeze dry.

2. Put the mushrooms into a steaming bowl. Add enough water to cover. Put in all the remaining ingredients and steam for 1 hour. Leave it to cool. Transfer the mushrooms into a storage box. They last well in the fridge for 1 to 2 weeks.

燒味・家常篇

Home-style recipes

自家製作燒味菜式，不難！

在傳統的燒味上加點烹調心思，

如桂花汁鐵板叉燒、腐乳紅燒鴿等，

吃出街外與別不同的燒味滷水。

川滷牛臉珠

Beef cheek in
Sichuan spiced marinade

燒味・家常篇

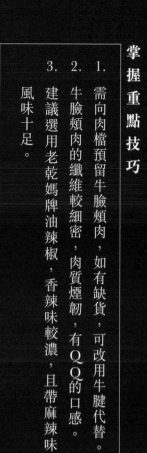

掌握重點技巧

1. 需向肉檔預留牛臉頰肉，如有缺貨，可改用牛腱代替。

2. 牛臉頰肉的纖維較細密，肉質煙韌，有QQ的口感。

3. 建議選用老乾媽牌油辣椒，香辣味較濃，且帶麻辣味，風味十足。

材料

牛臉頰肉	600 克

川味滷水料

生油	4 湯匙
薑片	25 克
乾葱茸	20 克
蒜茸	20 克
紅長尖椒	1 隻
川椒粒	2 湯匙
八角	1 粒
豆瓣醬	1 湯匙
油辣椒	1 湯匙

滷水調味料

清水	300 克
生抽	150 克
片糖	150 克
幼鹽	半茶匙
五香粉	1 茶匙
雞粉	半湯匙
唐芹	40 克
葱	15 克
花椒辣油（後下）	1 湯匙
花雕酒（後下）	1 湯匙

做法

1. 將牛臉頰肉洗淨，放入滾水內待翻滾，轉慢火煮 1 小時，取出，放入冰水內降溫，瀝乾備用。

2. 燒熱油鑊，先爆香川味滷水料，再加入滷水調味料，煮滾後放入牛臉頰肉轉慢火煮 15 分鐘，熄火焗 15 分鐘。

3. 待涼後，放入冰箱，食用時切片，最後澆上滷水汁即成。

白胡椒豬手

Pork trotter in white pepper stock

材料
急凍豬手件	12 件
薑	10 克
葱	10 克

滷水味料
清水	1.5 公斤
八角	1 粒
唐芹	40 克
薑	20 克
葱	20 克
香葉	3 片
白胡椒粒	15 克
幼鹽	40 克
雞粉	20 克

胡椒湯味料
清水	1.2 公斤
白胡椒粒	25 克
八角	1 粒
香葉	2 片
幼鹽	10 克
雞粉	10 克

掌握重點技巧

1. 建議用金福牌白胡椒粒，辛辣味道較濃香。

2. 煮腍的豬手浸在滷水味料時，需貯存於雪櫃，進食時取出，用胡椒湯味料翻煮即可。

3. 若購買原隻豬手，必須斬成八小件再烹調。

做法

1. 將兩份白胡椒粒分別炒香，再用刀壓成碎粒，備用。

2. 滷水味料煮滾，轉文火煲 10 分鐘，關火待涼。

3. 胡椒湯味料煮滾（白胡椒碎用煲湯袋盛起），轉慢火煲 10 分鐘，關火備用。

4. 豬手解凍後，燒掉細毛，剪去趾甲，洗淨。

5. 燒開水，放入豬手、薑及葱，翻滾後，轉慢火煲 1 小時 30 分鐘，待豬手脤身，取出，用清水沖洗至豬皮涼爽，瀝乾水分。

6. 滷水味料過濾後，放入豬手浸醃約 4 小時。

7. 享用時，煮滾胡椒湯味料，放入豬手，用慢火煮 3 分鐘，上碟。

燒味・家常篇

千層東坡塔
Dongpo pork pagoda

燒味・家常篇

掌握重點技巧

1. 五花腩肉燜至腍身，並急凍至硬身才容易切成薄片。

2. 建議選用李錦記金醬油，其味道與一般生抽不同，鹹味較淡且帶鮮味。

3. 塔型模具在上海街的廚具店舖有售，型號是 H28 蛋糕模。

材料

去骨五花腩	600	克
甜梅菜	300	克
生油	2	湯匙
乾葱片	50	克

滷水材料

薑	20	克
葱	20	克
蒜肉	20	克
乾葱肉	20	克
八角	3	粒

滷水調味料

清水	750	克
片糖	45	克
金醬油	230	克
花雕酒	230	克
幼鹽	1	茶匙
雞粉	2	茶匙
五香粉	1	茶匙
丁香	10	粒

芡汁調味料

花雕酒	10	湯匙
滷水汁	500	克
清水	300	克

芡汁

生粉	1	湯匙
清水	1	湯匙

做法

1. 五花腩放入沸水內，飛水 5 分鐘，盛起，沖冷水洗淨，備用。

2. 滷水汁製法：用生油 2 湯匙起鑊，爆香滷水材料後，加入調味料煮滾；煲底鋪上竹墊（以免材料燒焦），放入五花腩，滾後轉慢火加蓋煲 2 小時，待腩肉腍身，取出冷卻，滷水汁隔渣留用。

3. 五花腩肉皮向下放在平盤上，放入速凍冰箱冷藏至硬身，再切成約 4mm 薄片備用。

4. 甜梅菜剪成小棵，用水浸 1 小時，洗淨幼砂，瀝乾水分，切碎。

5. 燒熱鑊，下生油 2 湯匙爆香乾葱片，下梅菜炒勻，加入芡汁調味料，加蓋慢火煮 15 分鐘，盛起梅菜，芡汁調味料留用。

6. 腩肉片交叉疊放，築成一層層的塔型，放入塔型模內，釀入梅菜（盡量填滿及壓緊），塔尖位向下放好，蒸 30 分鐘，取出及脫模，煮滾芡汁調味料後，加入預先混和的芡汁勾芡，澆上東坡肉塔即成。

腐乳紅燒乳鴿

Deep fried squabs with
fermented beancurd sauce

掌握重點技巧

1. 如選用焗爐烘乾乳鴿，小心爐溫過熱，令鴿表皮的麥芽糖皮溶化及燒焦。爐溫宜約攝氏一百度，而且將爐門半張開。

2. 當乳鴿風乾或烘乾後，表皮以乾爽及不黏手為佳。

3. 蘸腐乳汁品嘗，令乳鴿更惹味，是一款新嘗試。

材料

頂鴿	2 隻（每隻約 480 克）	

滷水材料

清水	3 公升	
薑片	40 克	
香茅	40 克	
八角	5 粒	
香葉	20 片	
丁香	2 湯匙	
桂皮	10 克	
甘草	10 片	
白胡椒粒	5 克	
沙薑片	20 片	
小茴香	4 湯匙	

滷水調味料

冰糖	340 克	
幼鹽	230 克	
白腐乳	230 克	
雞粉	40 克	
花雕酒	40 克	
玫瑰露酒	40 克	

上皮料

A 類上皮水	約 50 克	
（做法參考 p.13）		

蘸汁（腐乳汁）

腐乳	3 件	
花雕酒	2 湯匙	
砂糖	1 湯匙	
麻油	1 茶匙	

* 拌勻

做法

1. 將滷水材料煮滾後，轉慢火熬 1 小時，加入滷水調味料，待冰糖煮溶後熄火。
2. 頂鴿剪腳，去油膏、肺及喉管，洗淨，放入已煮滾的滷水料中，待翻滾後轉慢火煮 5 分鐘，熄火焗 25 分鐘。
3. 取出鴿，用熱水沖走表皮油脂，用上皮水塗勻鴿皮，用風筒吹乾或放入焗爐烘乾。
4. 燒熱油，將乳鴿炸至外皮香脆，伴蘸汁享用。

燒味 · 家常篇

Barbecue pork on hot plate with osmanthus molasses

桂花汁鐵板叉燒

材料

梅頭肉	約 1 公斤
洋葱	1 個
百花蜜	1 瓶（燒烤用）
生粉	8 湯匙

醃料

砂糖	5 湯匙
幼鹽	1 湯匙
雞粉	半湯匙
海鮮醬	2 湯匙
蠔油	2 湯匙
芝麻醬	半湯匙
生抽	半茶匙
老抽	半茶匙
乾葱茸	半湯匙
蒜茸	1 茶匙
玫瑰露酒	半湯匙
雞蛋	1 隻

桂花汁

清水	60 克
片糖	40 克
生抽	20 克
百花蜜	120 克
鹽	10 克
乾桂花	1 茶匙

燒味・家常篇

掌握重點技巧

1. 梅頭肉用生粉拌勻時，必須加入少許清水令表面形成一層薄糊狀，才可發揮滲透的作用。

2. 梅頭肉用生粉醃四小時，具清潔及去掉血水的作用，也令梅頭肉鬆軟及嫩滑。

3. 焗爐必須預熱足夠溫度才放入梅頭肉烘焗，否則肉面的汁料不能在十分鐘內沸騰，影響味道。

4. 梅頭肉先用高溫（攝氏二百四十度）烘焗，再轉低溫（攝氏一百度）慢烤，可封鎖肉汁，而且更美味。

做法

1. 梅頭肉切成約 11 吋長 x 2.5 吋寬 x 0.8 吋厚長方形，用水沖洗 30 分鐘，盛起，放入生粉 4 湯匙拌勻醃 4 小時，放於雪櫃保鮮。

2. 用水沖去梅頭肉表面生粉與血水，再放入生粉 4 湯匙拌勻。將醃料拌勻，放入梅頭肉再攪拌醃 40 分鐘。

3. 桂花汁煮滾，備用。

4. 預熱焗爐，梅頭肉平放在烘架上，放於用錫紙包好的底盤上，用高溫（240℃）烘焗 10 分鐘，翻轉另一面，烘焗 10 分鐘至表面微焦黑，轉低溫（100℃）烘 30 分鐘，取出塗百花蜜。

5. 焗爐再調至高溫（240℃），放入梅頭肉烘至兩面的百花蜜沸騰，轉用低溫（100℃）再烘 30 分鐘，取出塗勻百花蜜，即成叉燒。

6. 燒熱鐵板，將叉燒斜切成薄片；洋葱切成粗粒狀，備用。

7. 取出已燒熱的鐵板，放上洋葱粒，排上叉燒片，最後澆上桂花汁即可。

紅油牛腱 Spicy soy-marinated beef shin

材料

急凍牛腱	1 條（約 500 克）

滷水材料

生油	2 湯匙
乾葱肉	20 克
蒜肉	10 克
薑片	20 克
唐芹	25 克
豆瓣醬	20 克
指天椒（切粒）	2 隻

滷水調味料

清水	1.2 公斤
生抽	160 克
鮮醬油	40 克
花雕酒	60 克
老抽	1 湯匙
沙茶醬	1 湯匙
蠔油	20 克
幼鹽	10 克
雞粉	20 克
五香粉	1 湯匙
辣椒粉	1 湯匙
片糖	160 克

做法

1. 牛腱解凍後，放入滾水煮 5 分鐘，取出用清水沖洗 5 分鐘，用圓針插（或餐叉、長竹籤）在牛腱上均勻地戳入，令牛腱容易入味。

2. 將滷水材料炒香，加入調味料煮滾，放入牛腱待滾後轉慢火，加蓋煮約 1 小時 30 分鐘，見牛腱軟腍，盛起待涼。

3. 將滷水調味料過濾，待冷卻後，放入已涼的牛腱，貯存於雪櫃。

4. 享用時，將牛腱切成薄片，澆上調味汁即成。

掌握重點技巧

1. 烹調滷水時，需攪拌至調味料完全溶解，以免糖份黏底煮焦。

2. 以慢火煮牛腱時，必須加蓋封着，令牛腱易腍，也使香氣及水分不容易散失。

燒味・家常篇

蜜汁鴨腳包

Duck feet rolls in molasses

材料

材料	分量	單位
炸鴨腳	6	隻
滷水豬肚	60	克
叉燒	60	克
榨菜	60	克
生雞肝	3	個
生雞腸	180	克
麥芽糖	40	克
叉燒醬	2	湯匙
牙籤	12	支

醃料

醃料	分量	單位
砂糖	9	湯匙
幼鹽	3	茶匙
雞粉	1.5	湯匙
生抽	1.5	湯匙
南乳（不連汁）	1.5	湯匙
南乳汁	1.5	湯匙
海鮮醬	3	湯匙
芝麻醬	2	湯匙
玫瑰露酒	2	湯匙
薑茸	1.5	湯匙
乾葱茸	3	湯匙
蒜茸	3	湯匙
五香粉	1.5	茶匙
生粉	2	茶匙
雞蛋	2	隻
花雕酒	2	湯匙

掌握重點技巧

1. 可用鵝腸或腩仔片代替雞腸。

2. 生雞腸的油膏多，用刀小心刮淨；浸熱水時勿太久，以免過度收縮令質感太韌。

3. 包紮鴨腳包時，將雞腸光滑一面向內，較深色及粗糙的向外，容易抹上叉燒醬的色澤。

4. 如雞腸的長度不足，可接上第二條再捲。

5. 包紮鴨腳包後，插入兩支牙籤成十字型以固定位置（避免插入鴨骨位）。

6. 建議選用同珍叉燒醬，增添鴨腳包表層的醬香味，而且燒烤後的色澤更鮮明。

即時睇片

做法

1. 炸鴨腳用熱開水煮滾後，轉慢火煲 30 分鐘，取出，放入清水內冷卻。

2. 生雞腸用刀刮去肥油，加入生粉 1 湯匙及幼鹽半湯匙拌匀，洗淨，再放入熱水內浸片刻至略收縮，取出沖水至冷卻，瀝乾水分備用。

3. 榨菜切成約 2.5 吋長筷子狀，用清水沖去鹹味，備用。

4. 豬肚及叉燒切成 2.5 吋長條狀；雞肝切開兩半，連同炸鴨腳放入醃料內醃 40 分鐘。取另一碗，放入雞腸醃味（以免與其他材料拌匀後難以取出）。

5. 將以上材料各取 1 件，放於鴨掌上，以雞腸緊緊捲實紮好，以牙籤固定位置，塗上叉燒醬醃 5 分鐘。

6. 預熱焗爐，放入鴨腳包用 240℃ 每面各烘 10 分鐘，抹上麥芽糖，再用 200℃ 每面各烘 5 分鐘，取出塗抹麥芽糖即成。

砂鍋豉油雞

Soy marinated chicken in casserole

材料

光雞	1 隻（約 2 斤 4 兩）	
玫瑰露酒	160	克
花雕酒	160	克
老抽	40	克

滷水料

清水	3	公升
生抽	3	公升
紅麴米	20	克
陳皮	1	個
甘草	10	克
香葉	10	克
桂皮	10	克
丁香	5	克
沙薑粒	20	克
小茴香	10	克
八角	5	粒

滷水調味料

冰糖	2	公斤
幼鹽	350	克
雞粉	100	克

雞汁味料

熱水	80	克
生抽	30	克
蠔油	25	克
砂糖	60	克
老抽	1	茶匙
雞粉	1	茶匙
幼鹽	1/4	茶匙

炒鍋味料

生油	2	茶匙
乾葱肉	5 粒（切半）	
蒜肉	5 粒（切半）	
葱段	20	克
薑	5	大片
花雕酒	1	湯匙

掌握重點技巧

1. 煮滷水的容器，最好選用較窄口的高身鋼煲，滷水汁的份量能夠將原隻雞浸過面，才容易令雞熟透及入味。

2. 滷水汁用後需要再次煮滾，待涼後貯存於雪櫃，留待日後使用。

3. 紅麴米在香料店有售。

做法

1. 將滷水料煮滾，轉慢火煲約 30 分鐘至香料散發，熄火，加入滷水調味料，攪動至冰糖完全溶解，即成豉油雞滷水汁。

2. 預先將雞汁味料煮滾，熄火備用。

3. 光雞取去內臟及肥油，洗淨。煮滾滷水汁，放入光雞待翻滾後，轉慢火滾 5 分鐘，熄火浸約 35 分鐘，至熟透取出，斬件，盛起。

4. 燒熱砂鍋，下油 2 茶匙，放入炒鍋味料炒香，潷入花雕酒，下豉油雞件，澆上雞汁味料，加蓋煮滾即成。

燒味・家常篇

燒味・宴客篇

Banquet recipes

京式片皮鴨、燒乳豬、潮州滷水鵝……

看似繁複的製作步驟，

只要預備功夫充足，

你可輕輕鬆鬆在家宴請親朋，

令親友讚好！

京式片皮鴨

Peking duck

掌握重點技巧

1. 將鴨放入油鑊後，見表皮出現氣泡，可用鵝尾針刺去氣泡，以免表皮漲起，賣相不美觀。

2. 鴨薄餅可在售賣粉麵的店舖訂購，毋須花時間製作。

材料

米鴨	1 隻（約 4 斤）
青瓜	200 克
大京葱	150 克
回味醬	4 湯匙
（做法參考 p.15）	
鴨薄餅（饃饃）	24 片

上皮料

B 類上皮水	150 克
（做法參考 p.14）	

醃料

砂糖	1.5 湯匙
幼鹽	1 湯匙
雞粉	1 茶匙
五香粉	1/4 茶匙
八角	3 粒
陳皮	1 個
花雕酒	2 湯匙

鴨薄餅料

低筋麵粉	250 克
80℃ 熱水	200 克
鹽	1/4 茶匙
生油	半茶匙

燒味·宴客篇

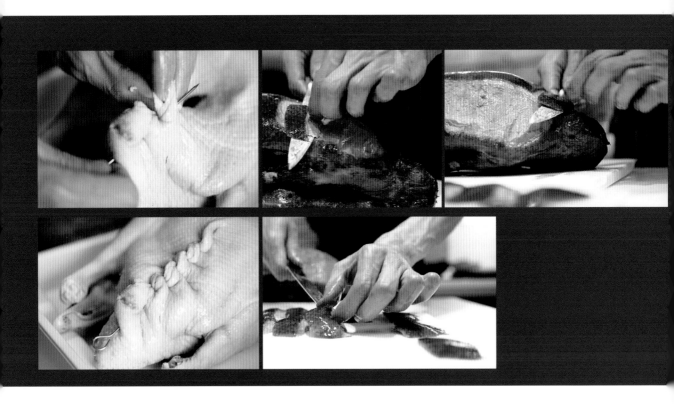

鴨薄餅做法

1. 熱水混和鹽及生油，倒入低筋麵粉內迅速攪勻，搓成光滑的麵糰，蓋上濕毛巾待涼。
2. 將麵糰搓成長條形，切成 24 小件，再將小件壓成圓薄片，放入鑊內用慢火煎熟定型，享用前再蒸 2 分鐘至熱即可。

做法

1. 米鴨去油膏、肺及喉管，洗淨，瀝乾水分。將醃料拌勻，倒入鴨腔內擦勻，用鵝尾針把尾部位置穿好。
2. 燒熱水，放入米鴨用滾水澆在鴨身，令表皮收緊及轉色，取出，用水冲洗至鴨皮涼爽，瀝乾水分。用鴨勾在兩旁鴨翼內側吊起，鴨頭向鴨背方向垂下，在鴨皮塗勻 B 類上皮水，吊起風乾備用。
3. 焗爐預熱至 150℃，鴨胸向上平放在焗盤，除去鵝尾針，將米鴨放進焗爐，烘 50 分鐘至表皮呈淺紅色（首 30 分鐘後先觀察鴨皮色澤，若見淡紅色，在鴨身面蓋上錫紙），再轉 100℃烘 30 分鐘至鴨熟透，取出吊起，瀝乾水分。
4. 鑊內燒熱生油，放入光鴨炸至皮脆及呈紅色，連皮帶肉斜片成鴨件。
5. 鴨薄餅（饃饃）蒸熱後，取出，放上京葱絲、青瓜條、回味醬及鴨件，包好享用。

葡汁焗羊排

Grilled lamb chops in
Portuguese sauce

掌握重點技巧

1. 醃料混和後，預早貯存一天，更散發香料味道，令香氣更濃。

2. 在步驟二，羊排用生粉醃三十分鐘，具清潔作用，而且形成一層薄糊，可鎖緊肉汁避免流失。

3. 嗜羊者可購買內蒙古的羊排，羶味較濃郁。

材料

羊排	6 件（每件 100 克）
生粉	12 湯匙
清水	8 湯匙
車厘茄	12 粒
紫洋葱絲	1 個

醃料

砂糖	10 茶匙
幼鹽	2 茶匙
雞粉	2 茶匙
孜然粉	1.5 茶匙
五香粉	半茶匙
黑胡椒粉	半茶匙
海鮮醬	2 茶匙
蠔油	2 茶匙
芝麻醬	1 茶匙
金醬油	2 茶匙
米酒	2 湯匙
乾葱茸	2 茶匙
蒜茸	2 茶匙
薑茸	2 茶匙

葡汁料

咖喱醬	1 湯匙
椰漿	4 湯匙
茄汁	1 湯匙
砂糖	1 茶匙
鹽	1 茶匙
雞粉	1 茶匙
牛油	20 克
清雞湯	2 杯
生粉	3 湯匙

做法

1. 醃料預早一天拌勻，貯存於雪櫃備用。

2. 羊排洗淨，用生粉 6 湯匙及清水 4 湯匙拌勻，形成一層薄糊狀待 30 分鐘，用水洗淨。

3. 羊排再用生粉水（生粉 6 湯匙及清水 4 湯匙）拌勻，使表面形成一層薄糊狀，然後與醃料拌勻醃 30 分鐘。

4. 燒熱油，放入羊排用中火煎至兩面金黃，盛起。

5. 焗盤內放入洋蔥絲半個，排上羊排及車厘茄。

6. 牛油起鑊，放入餘下的洋蔥絲炒香，下葡汁料拌勻煮熱，倒入生粉水煮滾，澆在羊排上。

7. 預熱焗爐 200℃，放入羊排焗至表面微焦黃色，即可享用。

燒味・宴客篇

南乳吊燒雞

Fried chicken in
red tarocurd sauce

掌握重點技巧

1. 雞身串上鐵針的工序，能夠將雞隻定型。

2. 醃料可用於相類似的菜餚，如南乳雞翼、乳鴿或鴨等，但醃味時間則視乎各道菜之需要。

3. 炸雞前用針刺雞皮，可去掉水分，以免炸雞時出現氣泡。

材料

光雞	1 隻（約 2 斤 4 兩）	

上皮料

A 類上皮水	120 克
（做法參考 p.13）	

醃料

南乳	680 克
生抽	680 克
白腐乳	80 克
花雕酒	80 克
乾葱肉	80 克
蒜肉	160 克
芫茜	40 克
葱	80 克
幼鹽	200 克
雞粉	120 克
薑	80 克
南薑	40 克
桂花陳酒	100 克

蘸料

生油	1 茶匙
南乳	80 克
清水	80 克
幼糖	60 克
雞粉	5 克
玫瑰露酒	5 克
花生醬	30 克

* 用生油起鑊，下已拌蘸料慢火煮滾。

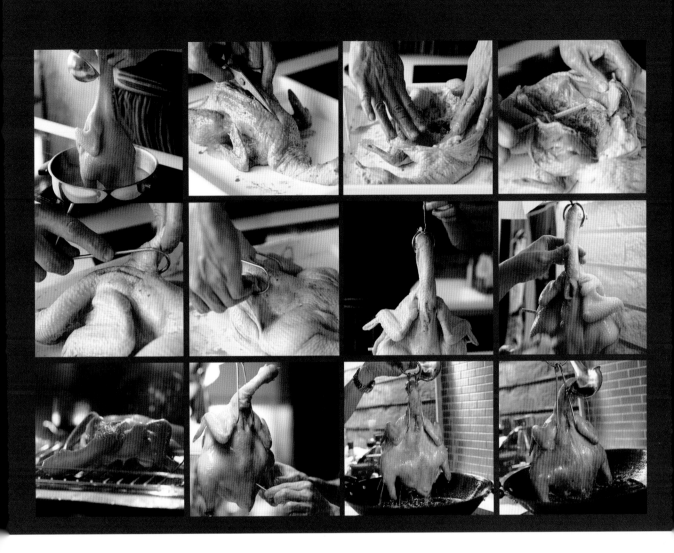

做法

1. 雞洗淨，放入滾水內燙至雞皮收緊，取出，用水沖洗至雞皮涼爽，備用。

2. 醃料用攪拌機打碎成醬汁，放入盤內，下雞隻醃 1 小時。

3. 雞隻從胸部中間剪開，拉開，串上鐵針（或用竹筷子代替），使雞身
   張開成琵琶形，用清水沖淨雞皮，瀝乾水分後，塗抹上皮水，風乾或
   用焗爐烘乾。

4. 焗爐預熱 150℃，放入雞（皮向上）烘 30 分鐘，轉 100℃再烘 30 分鐘，
   吊起，用針在雞皮刺數下。

5. 燒熱油，澆上油炸脆雞皮，伴已煮滾的蘸料享用。

燒味・宴客篇

55

潮州滷水鵝

Spice marinated goose
in Chaozhou style

材料
黑棕鵝	1 隻

潮州滷水料 A
鵝油	10 湯匙
蒜肉片	160 克
乾葱肉片	160 克
薑片	160 克
葱段	160 克
花雕酒	200 克

潮州滷水料 B
清水	8 公升
八角	12 粒
陳皮	4 個
川椒	4 湯匙
五香粉	2 湯匙
南薑	320 克
芫茜	240 克
金不換	80 克
紅麴米	2 湯匙
金華火腿	160 克

滷水調味料
生抽	1.6 公斤
老抽	320 克
幼鹽	120 克
片糖	1.2 公斤
雞粉	160 克
蠔油	1.2 公斤

蒜香醋汁
潮汕米醋	8 湯匙
砂糖	4 湯匙
幼鹽	半茶匙
蒜茸	2 茶匙
紅椒粒	1 茶匙

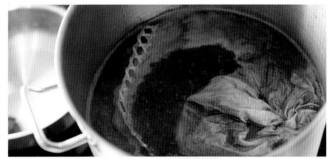

做法

1. 蒜香醋汁做法：米醋、砂糖及幼鹽攪拌至完全溶化，加入蒜茸及紅椒粒拌勻，備用。

2. 光鵝去掉油膏（留作製鵝油之用）及內臟，洗淨，瀝乾水分，剪開頸部氣管位及肚腩位多餘皮脂，備用。

3. 鵝油膏放入白鑊內，用慢火熬製成鵝油，留作滷水用途。

4. 爆香滷水料 A 留用，令滷水汁更香濃。

5. 滷水料 B 煮滾，轉慢火煲約 1 小時，熄火，加入滷水調味料至完全溶解，倒入預先製作的滷水料 A 拌勻，即成潮州滷水汁。

6. 滷水汁煮至微滾，放入光鵝，待滾後轉慢火煲約 1 小時，至鵝身全熟，取出待涼。

7. 享用時，將滷水鵝的頭頸及背斬件，其他部位去骨後，切成鵝肉片，上碟，澆上熱滷水汁，蘸蒜香醋汁享用。

燒味・宴客篇

8　　6　　3　　1

9　　7　　4　　2

10　　5

脆皮琵琶鵪鶉

Crispy "Pipa" quails

材料

鵪鶉	3 隻
回味醬	3 湯匙
（做法參考 p.15）	
竹籤（約 6 吋）	3 枝

上皮料

B 類上皮水	3 湯匙
（做法參考 p.14）	

醃料

生抽	1.5 公斤
清水	300 克
雞粉	60 克
幼鹽	20 克
薑片	60 克
陳皮	2 個
甘草片	6 片
沙薑	6 粒
香葉	7 片
八角	4 粒
桂皮	20 克
南薑片	20 克
桂花陳酒（後下）	120 克

掌握重點技巧

1. 在鵪鶉皮鬆針，即用鵝尾針在表皮均勻地刺入，在炸製時令表皮的空氣流通，不易出現氣泡。

2. 在鵪鶉內腔穿上竹籤使其定型，不易收縮，可做成琵琶形狀（參考第五十五頁「南乳吊燒雞」）。

燒味・宴客篇

61

做法

1. 醃料用中火煮滾，轉慢火煲 10 鐘，熄火待涼。

2. 鵪鶉去內臟、洗淨，用滾水澆在表皮至收緊及轉色後（由白轉成淡黃），用清水沖洗至表皮涼爽，放入醃料內浸醃 30 分鐘。

3. 鵪鶉由胸部中間剪開，把鵪鶉身壓平，在腔內中央肋骨位穿上竹籤定型，沖淨表皮，抹乾水分，在表皮塗勻上皮水，風乾或用焗爐以低溫（50℃）烘乾。

4. 預熱焗爐至 150℃，鵪鶉平放烘架上（皮向下），內腔向上塗勻回味醬，烘 5 分鐘後，將鵪鶉反轉（皮向上），改用 100℃烘 30 分鐘取出。

5. 在鵪鶉表皮鬆針後，放入滾油內將表皮炸脆，盛起，在內腔塗勻回味醬，斬件即成。

燒味·宴客篇

瑤柱金腿貴妃雞

Concubine chicken with
dried scallops and ham

材料

光雞	1 隻（約 2 斤 4 兩）

瑤柱滷水料

清水	3.6 公升
大地魚乾	60 克
（烘乾，撕出魚肉）	
金華火腿	120 克
蝦米	120 克
乾瑤柱	150 克
薑肉	120 克
南薑	60 克
沙薑片	10 片
小茴香	2 湯匙
丁香	半湯匙
香葉	10 片
八角	3 粒
豬湯骨	300 克

滷水調味料

片糖	600 克
幼鹽	350 克
雞粉	60 克
紹酒	45 克
桂花陳酒	30 克
玫瑰露酒	15 克

蘸雞料

薑茸	2 湯匙
葱茸	2 湯匙
幼鹽	1/4 茶匙
雞粉	1/4 茶匙
熱油	3 湯匙

掌握重點技巧

1. 瑤柱滷水料中，可用大地魚末三十克代替大地魚乾，使用時較方便。

2. 浸雞的滷水汁味道較濃，浸焗約一小時即可取出，以免味道太鹹。

做法

1. 乾瑤柱用熱水浸軟；金華火腿洗淨，與大地魚乾及蝦米一起放入焗爐略烘至香味散出。將瑤柱滷水料（豬湯骨除外）用煲魚湯袋盛好。

2. 豬湯骨飛水，與煲魚湯袋的香料放入清水內煮滾，轉慢火煮 2 小時，熄火。

3. 將調味料加入瑤柱滷水內，待片糖溶解後放涼，棄掉湯骨，滷水汁留用。

4. 光雞去掉油脂、肺部及喉管，洗淨，放入滾水內，煮滾後轉慢火煲 5 分鐘，熄火浸焗 30 分鐘，取出，放入冰開水中冷卻。

5. 將雞取出，瀝乾水分，浸泡瑤柱滷水內約 1 小時，斬件上碟。

燒味‧宴客篇

香草汁燒乳豬

Roast suckling pig with rosemary

材料

急凍乳豬	半隻（約 1.6 公斤）
鍋巴	12 塊
青瓜	40 克
食用梳打粉	半茶匙

醃料

砂糖	3.5 湯匙
幼鹽	2 湯匙
五香粉	1/4 茶匙

上皮料

A 類上皮水	4 湯匙
（做法參考 p.13）	

香草汁料

清水	80 克
巴黎牛油粉	16 克
雞粉	3 克
砂糖	3 克
迷迭香	1 克

即時睇片

做法

1. 乳豬解凍，背向下平放，將豬下巴斬開，切去喉核及喉管，挖去豬腦及豬身兩側的網油。

2. 由豬尾脊骨處從尾至頭劈開（只劈開脊髓骨而不弄穿外皮），用雙手按下令乳豬身攤開。

3. 切去近頭部的前四條排骨；慢慢沿關節位起出前膊骨及琵琶骨，在厚肉位置劃數刀；用刀片起後髀肉，在厚肉部位劃數刀，最後斬去豬手、豬腳及豬尾，用勾掛起乳豬。

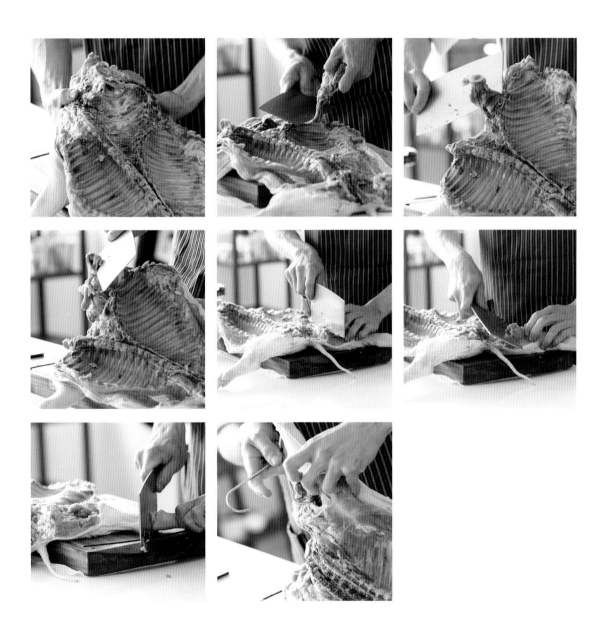

即時睇片

4. 燒開水，放入乳豬用中火焓至豬皮收緊（約 2 分鐘），再用清水冲至豬皮涼爽，瀝乾水分。

5. 乳豬平放盤內（皮向下），將醃料混合，塗抹於豬腔內的肉（別搽在豬皮上）醃 10 分鐘，讓盤子傾斜，令豬腔內的多餘水分流走，不影響烤烘後外皮的質素。

6. 乳豬皮向上平放，用幼鹽搽勻豬皮表面待 2 分鐘。將半茶匙食用梳打粉加入 600ml 溫水（約 70℃）內拌勻，洗擦乾淨豬皮的鹽分及油脂，用水再冲洗，用毛巾抹去水分，在豬皮上均勻地搽勻上皮料。

7. 預熱焗爐，乳豬平放烘架（背向上），用錫紙包好底盤，放入焗爐內（爐門必需保持半開）用 250℃ 烘 15 分鐘，再轉 150℃ 烘 15 分鐘，最後用 100℃ 烘約 1 小時至乳豬肉熟透，取出風乾。

8. 香草汁料煮至稀糊狀，熄火備用；青瓜切成圓片。

9. 燒熱油（約 180℃），放入鍋巴炸至膨脹及香脆，瀝乾油分。

10. 用鵝尾針在乳豬皮平均刺入數下（稱為鬆針），放入滾油內炸至皮脆，起皮、去肥膏，肉層去骨，切成 1.5 吋 x 1.5 吋的正方形。

11. 享用時，鍋巴放於底層，鋪上青瓜片及乳豬肉，澆上半湯匙香草汁，最後放上乳豬脆皮即成。

掌握重點技巧

1. 處理乳豬時，在厚肉部位劃數刀，能夠加速醃味的效果。

2. 塗抹上皮料時，必須由上至下連續掃抹，令上皮料平均地塗在豬皮上。

3. 因應家中焗爐爐大小，切去豬頭或將乳豬切開烘焗；豬頭可留作滷水之用。

4. 炸鍋巴時油溫要稍高，才可炸至鬆脆的口感。

燒味・海鮮篇

Seafood recipes

打破一般舊有的概念，

利用製作燒味滷水的方法炮製海鮮，

卻保留海鮮的鮮甜美！

為燒味賦上新意念、新方向。

回味醬焗鮮鮑魚

Braised abalones in
unforgettable sauce

材料

鮮鮑魚仔	12 隻
（每隻約 45 克）	
回味醬	200 克
（做法參考 p.15）	

味料

清水	170 克
幼鹽	1 茶匙
砂糖	半茶匙
雞粉	半茶匙
玫瑰露酒	2 茶匙
薑片	30 克
乾葱茸	2 湯匙
蒜茸	2 湯匙

燒味・海鮮篇

做法

1. 將鮮鮑魚肉表面刷淨，取出鮑魚肉，去除內臟洗淨，外殼用清水煮熱，洗淨留用。

2. 鮑魚肉灑上生粉 2 湯匙略擦約 2 分鐘（放於笪箕內較佳），用清水洗淨。

3. 燒滾味料之清水，放入薑片用慢火煮 5 分鐘，待冷卻，加入其他配料拌勻，放入鮑魚肉浸泡 30 分鐘，抹乾水分。

4. 鮑魚肉用回味醬拌勻醃 30 分鐘。

5. 焗盤包上錫紙，預熱焗爐 240℃，放入鮑魚每面各烘 4 分鐘即成，享用時排放外殼內上碟。

三杯醬焗花蟹

Three-cup crabs

掌握重點技巧

1. 緊記蟹蓋毋須泡油，因蟹蓋太薄容易爆裂，只需在慢火炒蟹時，放於蟹件上，加蓋焗熟即可。

2. 若選用肉蟹或膏蟹，烹煮時間要多加二至三分鐘，因蟹殼較厚，難以煮熟。

材料

花蟹	2 隻（約 600 克）
生粉	4 湯匙
麻油	2 湯匙

料頭

葱	30 克
薑	20 克
蒜肉	16 粒
紅辣椒	2 隻
九層塔	15 克

三杯醬料

清水	1 杯
砂糖	1 茶匙
老抽	1 茶匙
生抽	1 茶匙
豆瓣醬	1 茶匙
蠔油	1 湯匙
米酒	2 湯匙
花雕酒	2 湯匙

* 拌勻

芡汁

清水	2 湯匙
生粉	1 湯匙

做法

1. 蟹腹朝上，用刀在中央破開，拉去外殼，取出內臟，洗淨、切塊，蟹鉗斬開兩件，略拍至有裂縫，均勻地灑上生粉。

2. 薑切片；葱及紅辣椒切段；九層塔分開莖及葉片。

3. 燒熱鑊下油，下薑片、蒜肉及花蟹用中油溫泡油至金黃色，瀝乾油分備用。

4. 燒熱麻油，放入料頭（九層塔除外）用中火炒香，下花蟹拌勻，倒入三杯醬料攪勻，加蓋，慢火煮至水分略收乾（約 5 分鐘），加入九層塔，下芡汁勾芡，拌炒至水分收乾，上碟享用。

燒味・海鮮篇

燒酒煮花螺

Babylon conches in
Chinese wine

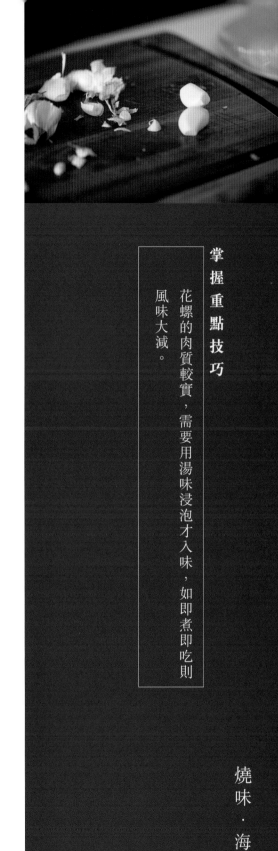

材料

花螺	600 克
清水	1.5 杯
生油	2 湯匙
芫茜段	20 克
葱段	20 克

料頭

薑茸	20 克
乾葱茸	20 克
蒜茸	20 克
紅辣椒	1 隻
川椒	1 湯匙

調味料

八角	2 粒
香葉	3 片
紫蘇葉	2 片
沙茶醬	1 湯匙
豆瓣醬	2 茶匙
花生醬	1.5 湯匙
金醬油	1 湯匙
老抽	1 湯匙
砂糖	2 茶匙
鹽	1 茶匙
雞粉	1 茶匙

燒酒料

玫瑰露酒	半湯匙
花雕酒	1.5 湯匙
米酒	1.5 湯匙

做法

1. 花螺洗淨，用開水煮約 2 分鐘，瀝乾水分備用。

2. 燒熱鑊，下生油 2 湯匙爆香料頭，放入花螺拌炒，下清水 1.5 杯及調味料慢火煮 5 分鐘。

3. 最後加入燒酒料煮滾，熄火，浸泡約 1 小時以上。食用時加入芫茜段及葱段煮熱，盛起上碟。

掌握重點技巧

花螺的肉質較實，需要用湯味浸泡才入味，如即煮即吃則風味大減。

燒味・海鮮篇

海蝦滷汁千張包

Spice-marinated Bai Ye dumplings
with shrimp and pork filling

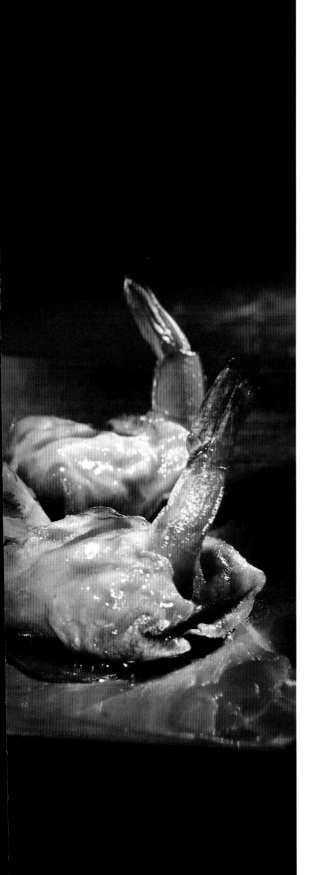

材料

百頁	3 張
（每張約 9 吋 x 9 吋）	
鮮蝦（留尾）	12 隻
椰菜（切碎）	90 克
食用梳打粉	半茶匙
溫水	2 杯

餡料

豬肉碎	100 克
葱花	5 克
薑茸	3 克
蒜茸	3 克

調味料

五香粉	1/8 茶匙
生油	1.5 茶匙
萬字醬油	1 茶匙
花雕酒	1 茶匙
砂糖	2/3 茶匙
幼鹽	1/4 茶匙
雞粉	半茶匙
麻油	1/4 茶匙

黏皮料

糯米粉	2.5 湯匙
熱水	3.5 湯匙

* 拌勻

滷汁料 A

生油	2 湯匙
紅尖椒	5 克
薑片	5 克
乾葱片	5 克
蒜片	5 克
八角	1 粒
川椒粒	1.5 湯匙
花雕酒	1 湯匙

滷汁料 B		
清水	450	克
生抽	1.5	湯匙
老抽	半	湯匙
紅糖	半	湯匙
幼鹽	半	茶匙
雞粉	半	茶匙
唐芹	5	克
葱	5	克
芫茜	5	克
五香粉	半	茶匙
花椒辣油	1	湯匙

芡汁		
生粉	1	湯匙
清水	2	湯匙

掌握重點技巧

將百頁浸泡梳打粉水，除令百頁的顏色變白外，也有軟化百頁質感的功效。

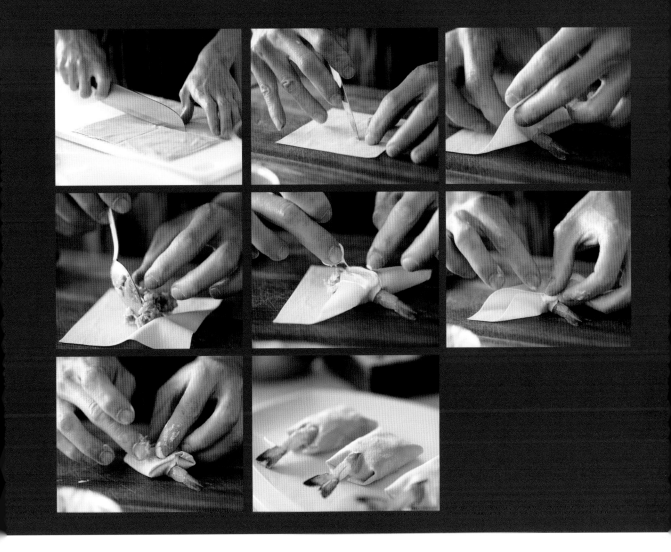

做法

1. 溫水 2 杯與食用梳打粉半茶匙拌溶，下百頁浸泡 15 分鐘至顏色變白，用清水漂洗百頁 15 分鐘。

2. 將餡料攪拌均勻，加入調味料用手攪勻至肉餡帶黏性，加入椰菜碎混合，冷藏 1 小時備用。

3. 百頁一張切成 4 小片，在百頁輕�}一刀，穿過蝦尾，在蝦身放上肉餡 1 湯匙，包成小包，用黏皮料封口。

4. 燒熱油鑊，用慢火將小百頁包煎至金黃色，取出備用。

5. 爆香滷汁料 A，加入滷汁料 B 煮滾，放入小百頁包，待翻滾後轉慢火，加蓋煮 5 分鐘。

6. 小百頁包放碟上排好，滷汁料埋芡，澆在小百頁包上即成。

燒味・海鮮篇

宮保芝士龍蝦焗烏冬麵

Lobster in
Kung Pao cheese sauce
with Udon noodles

掌握重點技巧

處理龍蝦時，宜先用木筷子從龍蝦尾部戳入放尿，以免龍蝦肉帶有異味。

材料

龍蝦	1 隻（約 900 克）
烏冬麵	2 個
煙肉（切粒）	4 片

宮保芝士汁

芝士片	8 片
牛油	60 克
麵粉	3 湯匙
砂糖	3 茶匙
雞粉	1 茶匙
豆瓣醬	1 湯匙
洋葱粒	半個
青、紅甜椒粒	共 100 克
清水	2 杯

燜烏冬料

清湯	1 杯
鹽	1/3 茶匙
宮保芝士汁	半杯

宮保芝士汁做法

1. 牛油用慢火煮溶，放入洋葱粒炒香，灑下麵粉拌勻成粉漿，慢慢加入清水拌勻粉漿。
2. 放入其他調味料及芝士片，以慢火煮溶後，加入甜椒粒拌勻即可。

做法

1. 龍蝦在腹部位置直切開邊，洗淨、切件，瀝乾水分，灑上生粉約 1 湯匙，放入鑊內下油煎香備用。
2. 烏冬麵用熱水灼散，盛起。鑊內加入燜烏冬料拌勻，下烏冬麵用中火燜至軟身（約 2 分鐘），轉放焗盤內。
3. 龍蝦放於烏冬面，將餘下的宮保芝士汁澆於龍蝦上，最後灑上煙肉粒。
4. 預熱焗爐 240℃，放入龍蝦焗約 10 分鐘至表面金黃色即可。

材料

富貴蝦 .... 2 隻
（即瀨尿蝦，每隻約 250 克）
花雕酒 .... 4 湯匙

滷味料

清水 .... 600 克（1 斤）
薑片 .... 15 克
葱段 .... 15 克
幼鹽 .... 2 茶匙
雞粉 .... 1 茶匙
五香粉 .... 半茶匙
花雕酒（後下） .... 5 湯匙

醉香汁

花雕酒 .... 1 湯匙
桂花陳酒 .... 1 湯匙
糖桂花 .... 半湯匙

醉香富貴蝦

Wine-marinated mantis shrimps

掌握重點技巧

先用花雕酒浸泡富貴蝦，目的是將蝦灌醉，令它自動放尿，同時使肉質爽脆及帶酒香味。

燒味・海鮮篇

做法

1. 將滷味料煮滾，轉文火煲 10 分鐘，熄火，待冷卻，最後加入花雕酒拌勻，放入雪櫃備用。

2. 富貴蝦洗淨，瀝乾水分，用花雕酒 4 湯匙將蝦浸泡 15 分鐘。

3. 預熱蒸籠，放入蝦隔水蒸約 4 分鐘，取出，放入已冷卻的滷味料內待涼。

4. 用剪刀在蝦背剪開，放回滷味料內，冷藏醃泡 1 小時。

5. 享用時，去掉蝦殼，蝦肉切件上碟，澆上醉香汁即可。

燒味・海鮮篇

脆皮鰻魚

Crispy honey-glazed eel

燒味・海鮮篇

掌握重點技巧

步驟一處理白鱔的做法，可請魚販代勞，省卻時間。

材料

去骨白鱔	300 克
食用梳打粉	半茶匙
蜜糖	2 湯匙

上皮料

A 類上皮水	20 毫升
（做法參考 p.13）	

醃料

砂糖	5	茶匙
幼鹽	1	茶匙
雞粉	1	茶匙
海鮮醬	2	茶匙
蠔油	2	茶匙
沙嗲醬	1	茶匙
老抽	1	茶匙
玫瑰露酒	1	茶匙
花雕酒	1	茶匙
乾蔥茸	2	茶匙
蒜茸	2	茶匙
薑絲	1	湯匙

做法

1. 白鱔放於熱水略浸片刻，取出，擦去表皮潺液，切去頭尾，由鱔身中間開邊成一大塊魚肉，起出脊骨及內臟，洗淨，瀝乾水分。

2. 白鱔皮向上平放，用圓針插鬆針，在表皮搽上食用梳打粉醃 5 分鐘，用清水沖淨。

3. 醃料拌勻，放入白鱔醃 40 分鐘，取出，用清水沖去表皮的醃料，以竹籤在鱔件較厚肉的位置穿過並勾起，用滾水澆在表皮上至收緊，抹乾水分，塗勻上皮料，風乾備用。

4. 燒熱油鑊至 8 成熱，轉慢火，放入鱔肉，先將鱔皮向下煎至 8 成熟，反轉再煎熟鱔肉及呈金黃色，在鱔皮塗上蜜糖，最後斜切上碟享用。

燒味・節慶篇

Festive favourites

中國人，最重視節慶。

在喜慶的日子，

將燒味滷水元素融入節日食品裏，

炮製令人意想不到的好意頭祝福菜，

滿有驚喜！

蒜香脆皮腩仔

Garlic-scented crispy roast pork

材料

去骨豬腩肉	2 斤
薑片	80 克
葱	80 克

醃料

幼鹽	3 湯匙
砂糖	2 湯匙
雞粉	1 湯匙
五香粉	半茶匙
炸蒜茸	2 湯匙
生蒜茸	1 湯匙

上皮料

白醋	1 湯匙
玫瑰露酒	半湯匙
幼鹽（後下）	1 湯匙

蒜香醬

黃芥醬	1 湯匙
回味醬	1 湯匙（做法參考 p.15）
炸蒜茸	1 茶匙

＊拌勻

燒味．節慶篇

掌握重點技巧

1. 烘烤一段時間後，腩仔表面如出現焦黑，可用小刀刮去焦黑部份，再烘焗至表面呈幼粒狀為最佳效果。

2. 若表面某些位置呈現粗粒狀，可能腩肉受熱不均，建議將腩肉表面接近上火發熱線，再烘至呈幼粒點為佳。

做法

1. 豬腩肉洗淨，連同薑片及葱放入滾水內（煲底放竹底笪），待水翻滾後轉文火煮 20 分鐘，取出用水冲泡至肉涼及皮爽，瀝乾備用。

2. 豬腩肉平放（肉向上），用刀在腩肉上相間 1 吋闊剕開，塗抹醃料，醃 1 小時 30 分鐘。

3. 反轉腩肉平放（皮向上），用刀刮淨表皮，再用圓針插在豬皮均勻地戳下（別穿過脂肪層），再塗抹上皮料醃 30 分鐘，刮去表皮鹽分。

4. 預熱焗爐 200℃，豬腩肉皮向上平放烘架上，連同包好錫紙的底盆放進焗爐，半開門烘 10 分鐘至豬皮乾爽，取出，用圓針插在厚皮位置再戳數下（別穿過脂肪層）。

5. 焗爐溫度保持 200℃，放入豬腩肉（皮向上）再烘 30 分鐘，取出，用錫紙包好腩肉四邊，只露出豬皮，再放入焗爐用 250℃烘約 30 分鐘至皮脆，斬件，伴蒜香醬享用。

New Year shredded salad
with abalone and jellyfish

鮑撈起

掌握重點技巧

1. 灼海蜇的水溫不能太熱，約攝氏八十度即可，否則海蜇嚴重收縮及膨脹度不足，影響口感。

2. 貴妃雞肉做法參考第六十五頁的「瑤柱金腿貴妃雞」。

材料

海蜇皮	200 克
燒鵝肉	100 克
貴妃雞肉	100 克
罐裝即食鮑魚仔	1 罐
西芹	100 克
甘筍	100 克
皺皮瓜	120 克
哈蜜瓜	120 克
即食粟米片	60 克

醃料

幼鹽	半茶匙
砂糖	1 茶匙
雞粉	半茶匙
麻油	2 茶匙

撈起醬汁

日式胡麻醬	80 克
泰國雞醬	80 克
麻油	2 茶匙

做法

1. 撈起醬汁拌勻，冷藏備用。

2. 海蜇皮掃去表面粗鹽，捲成筒狀，切絲（切成筷子般厚度）。

3. 燒熱水至 80℃，放入海蜇絲浸至收縮即盛起，用清水沖至脹身及沒有鹹味（約 2 小時），瀝乾水分，冷藏貯存備用。

4. 燒鵝肉及雞肉切絲；鮑魚切片；皺皮瓜及哈蜜瓜去皮、切條。

5. 西芹及甘筍洗淨，切條，放入熱水內略灼，盛起，用冰水沖至涼，備用。

6. 海蜇與醃料拌勻，醃約 10 分鐘，隔水備用。

7. 海蜇排放於碟中央，其他材料分類伴於碟邊，鮑魚片放於頂部，澆上撈起醬汁，一齊舉筷撈起品嘗。

八寶釀鯪魚

Eight-treasure stuffed dace

材料

鯪魚　　　　　　　　　　　　　　　1 條
　（約 400 克，起肉、保留魚皮）

料頭

葱粒、冬菇粒、蒜茸　　　　　各適量

魚膠八寶料

鯪魚肉	約 400 克
蝦米	30 克
臘肉	30 克
臘腸	20 克
膶腸	20 克
冬菇	30 克
光茜·葱·陳皮	共 20 克

魚膠調味料

鹽	1 茶匙
雞粉	半茶匙
砂糖	1 茶匙
麻油	1 茶匙
生粉	2 湯匙
清水	4 湯匙
胡椒粉	少許

芡汁調味料

鹽	1/3 茶匙
蠔油	1 湯匙
老抽	1 茶匙
生粉	2 茶匙
清湯	半杯

掌握重點技巧

1. 將鮮鯪魚起肉、留皮，一般魚檔可代勞，只要說明用作釀鯪魚即可。

2. 起出的鯪魚肉不足夠釀入魚皮中，因此要另購魚肉補充，或加入已調味之鯪魚膠代替（若用此，魚膠的調味料份量需減少半份）。

3. 將鯪魚餡料冷藏半小時，令餡料更容易釀入鯪魚皮內。

4. 在鯪魚內腔塗上生粉，令餡料黏實魚皮，切件時不會分離，而且餡料釀入魚頭魚尾時要釀多一點，並按實，餡料不易脱落。

5. 鯪魚釀入餡料後，或許有點變形，需要用手捏成魚形。

6. 即釀即炸的效果最理想；餡料可預早一天拌妥，冷藏備用。

做法

1. 將臘味材料蒸 6 分鐘，取出，切成幼粒。

2. 冬菇、蝦米及陳皮浸軟，與芫茜及葱切成幼粒。

3. 大窩內放入鯪魚肉攪拌，搓至微微呈膠質，加入調味料順一方向拌勻，再用力搓擦至魚肉起黏性，拌至膠質，用力撻十數下成魚膠。

4. 八寶料放入魚膠內攪勻，冷藏半小時。

5. 在鯪魚皮內腔灑上生粉，釀入魚膠至整條魚緊實成魚形，在魚身灑上生粉。

6. 燒熱油，用中油溫炸釀鯪魚至半熟，盛起，再蒸約 15 分鐘至全熟，切件，排放碟上。

7. 爆香料頭，放入清湯及芡汁調味料勾芡，扒在釀鯪魚上享用。

香蕉脆魚塊

Fish toasts with banana

掌握重點技巧

1. 若不使用焗爐，可用油炸方式代替，但炸製時必須注意油溫，下材料時要用中火油溫（攝氏二百度），炸約四分鐘至表面金黃色即可，否則油溫太低，油分容易滲入方包內，令口感油膩。

2. 進食時，可用沙律醬或糖醋汁蘸吃，增加風味。

材料

魚柳	3 塊
方包	2 塊
香蕉	1 隻
火腿片	1 塊
牛油	30 克
芫茜	2 棵

醃料

鹽	半茶匙
雞粉	1/3 茶匙
麻油及胡椒粉	各少許
生粉	2 茶匙

蛋漿料

蛋液	2 隻
生粉	8 湯匙

做法

1. 魚柳及方包切成 1 吋 x 2 吋的長方形，魚塊用醃料醃勻。

2. 蛋液及生粉拌成蛋漿，備用。

3. 香蕉及火腿片切成小塊，加入少許蛋漿拌勻。

4. 方包底部塗上少許牛油，放上香蕉片，再鋪上蘸滿蛋漿的魚塊，加上火腿小片及芫茜，排放焗盤上備用。

5. 預熱焗爐 200℃ 約 10 分鐘，放入魚塊烤焗約 10 分鐘至方包鬆脆及呈金黃色即成。

燒味‧節慶篇

香酥叉燒月

Puff pastry mooncake with barbecue pork filling

水油皮材料

低筋麵粉	150 克
固體油	40 克
砂糖	20 克
雞蛋	半隻
吉士粉	20 克
清水	70 毫升

油酥皮材料

低筋麵粉	80 克
固體油	70 克

水油皮做法

1. 砂糖、雞蛋及固體油搓勻，逐少加入麵粉及吉士粉拌勻蛋漿，每次少份量加入水分，搓至麵糰光滑。
2. 將麵糰壓成長方形，放於長盤內用保鮮紙包好，靜待 1 小時後，再冷藏 1 小時即成。

油酥皮做法

麵粉與固體油搓勻至軟滑，壓成長方形放於盤內，用保鮮紙包好冷藏 2 小時備用。

月餅酥皮做法

1. 將水油皮用擀麵棒壓至油酥皮的 3 倍寬度。
2. 水油皮鋪於桌面，油酥皮放於中間，水油皮兩邊摺入包住油酥皮，用擀麵棒壓薄，對摺 3 層，再壓平至長方形，冷藏 1 小時。
3. 取出酥皮，對摺 3 層，用擀麵棒壓平至長方形，用保鮮紙包好冷藏 2 小時即可。

掌握重點技巧

1. 製作酥皮時，每次壓平至長方形後，必須冷藏至少一小時，否則壓皮時固體油容易溶化，使酥皮太濕潤而未能成型。

2. 包餡料時，盡量避免弄破餅皮，否則焗製時餅面容易爆裂。

燒味．節慶篇

叉燒月材料

叉燒	200 克
洋葱	50 克
鹹蛋黃	2 個（每個切成 4 份）
白芝麻	10 克
百花蜜糖	2 湯匙
雞蛋（拂匀）	1 隻

調味料

蠔油	2 湯匙
生抽	1 湯匙
老抽	1 湯匙
砂糖	4 湯匙

芡汁

水	50 毫升
粟粉	60 克

* 拌匀

叉燒餡做法

1. 叉燒及洋葱切幼粒，備用。
2. 燒熱鑊，下生油 3 湯匙炒香洋葱及叉燒，倒入清水 200 毫升煮熱，加入調味料煮溶，勾芡，煮至材料熟透，待涼後冷藏 1 小時備用。

叉燒月餅做法

1. 預熱焗爐 200℃ 約 10 分鐘。
2. 酥皮取出，用擀麵棒壓薄至 1/2cm 厚度，用 3 吋直徑圓模壓出圓餅皮。
3. 放入叉燒餡 1 湯匙及鹹蛋黃 1/4 個，收口包緊，搓成棋子型，排於焗盤上，在餅面掃上薄蛋液。
4. 放入焗爐焗約 15 分鐘，取出，掃上蛋液及灑上白芝麻；焗爐調低至 180℃，再焗約 5 分鐘，熄火，待 2 分鐘取出，塗上百花蜜糖，待涼享用。

Grilled spareribs in Peking sauce

京醬烤肉排

掌握重點技巧

1. 肉排放入焗爐前，建議澆上燜肉排汁，以免肉排表面太乾及容易燒焦。

2. 最後將燜肉排汁用中火煮滾至略稠，令醬汁的味道更濃郁。

3. 進食時伴金銀饅頭（蒸饅頭及炸饅頭）享用，蘸上香甜的肉汁，味道更可口。

材料

一字排	1 件（約 900 克）
京 葱	1 棵（切片）
洋 葱	1 個（切絲）

調味料

清 水	8 杯
茄 汁	4 湯匙
ok 汁	2 湯匙
唸 汁	1 湯匙
香 醋	2 湯匙
鮮 醬 油	3 湯匙
蠔 油	4 湯匙
鹽	2 茶匙
雞 粉	2 茶匙
冰 糖	200 克
紅 麴 米	20 克
薑 葱	50 克
香 料（草果、香葉、八角）	共 40 克

做法

1. 肉排塗上生抽上色，燒熱油 1 湯匙，下肉排用中火煎香。

2. 肉排放入煲內，加入調味料用中慢火煲約 1 小時 30 分至 2 小時至腍身。

3. 焗盤放入洋蔥絲，肉排去掉肋骨，排放洋蔥絲上，加入步驟 2 燜肉排汁一杯，放入焗爐用 220℃ 烤 15 分鐘，轉放碟內。

4. 燒熱步驟 2 的燜肉排汁，勾芡，澆於肉排上，以炒香的京葱裝飾即成。

端陽燒肉糉

Glutinous rice dumplings with
marinated pork filling

1. 滷水汁與糯米拌勻，令燒肉糉子色澤漂亮外，也充滿滷汁香。

2. 用蒸熟的糯米包糉，可縮短焗糉的時間。

3. 鹹蛋黃用焗爐烘香，令鹹蛋黃更甘香可口。

4. 包糉時，注意糉葉的摺疊位置，方向正確才能成功包好，以免糯米溢出。

5. 將包糉的線繩縛在釘子上，掛着包紮糉子，更方便！

即時睇片

材料

糯米	600 克
五花腩	150 克
眉豆	80 克
鹹蛋黃	5 個
栗子	10 粒
即食鮑魚（24 頭）	10 隻
冬菇	10 朵
米酒	1 湯匙
糉葉	30 張
線繩（每條一米長）	10 條

滷汁料 A

生油（先下）	4 湯匙
薑片	15 克
紅葱頭	80 克
八角	2 粒
花雕酒（後下）	1 湯匙

滷汁料 B

清水	600 克
生抽	1.5 湯匙
老抽	1 湯匙
紅糖	1 湯匙
幼鹽	半湯匙
蠔油	1 湯匙
雞粉	半湯匙
五香粉	半茶匙
香葉	4 片

燒味・節慶篇

做法

1. 冬菇用清水浸軟、蒸熟（做法參考 p.17）。
2. 乾糭葉用清水泡至稍軟，洗刷乾淨，放入滾水內煮至轉成碧綠色，取出，浸於冷水內備用。
3. 糯米用清水浸泡 5 小時，瀝乾水分，隔水蒸 30 分鐘，取出，包好備用。
4. 眉豆用清水浸泡 5 小時，瀝乾水分；栗子去殼、去衣，洗淨。
5. 鹹蛋黃用米酒 1 湯匙浸醃 5 分鐘，放入焗爐用 150℃ 烘 5 分鐘，待冷，切半。
6. 五花腩切成 10 件，每件重約 15 克。
7. 燒熱油鑊，下滷汁料 A 炒香，加入滷汁料 B 煮滾。滷水汁煲內放入竹底笪一塊，下冬菇、眉豆、栗子及五花腩件，翻滾後轉慢火加蓋煮 45 分鐘，取出材料備用。
8. 熟糯米盛於大盆內，加入滷汁 200 克及熟眉豆拌勻，分成 10 等份。
9. 包糭時，先取 2 張糭葉平放，半邊重疊（光滑面向上），反起糭葉尾，與葉頭對摺，再將其一邊沿摺入封口（約 1 吋闊），用手握緊封口位，形成半開口袋形，放入半份熟糯米，排上冬菇、鹹蛋黃、栗子、五花腩及鮑魚，鋪入半份熟糯米。
10. 取糭葉一張，包着半開袋口一邊，預留約 1 吋闊邊，向內摺入封口，將糭葉尖位向上，反摺向下，抓緊用線繩紮緊，隔水蒸 40 分鐘即成。

即時睇片

燒味・節慶篇

Glutinous rice balls with pork and sesame filling

元宵芝麻鹹肉湯圓

湯圓皮材料

糯米粉	350 克
溫水（約 40℃）	250 克
生油	1.5 湯匙
清水	2 湯匙

湯料

清雞湯	2 杯
清水	2 杯
幼鹽	1 茶匙
薑片	15 克

餡料

生油	1 湯匙
紅葱頭	40 克
豬頸肉	120 克
貢菜	50 克
冬菇	50 克
花雕酒	1 湯匙
老抽	1 茶匙
蠔油	1 茶匙
白芝麻	2 湯匙

芡汁

清水	2 湯匙
生粉	2 茶匙

燒味·節慶篇

湯圓皮做法

1. 大窩內放入糯米粉 4 湯匙，加入清水 2 湯匙揉成麵糰，放入滾水內煮至浮起成熟粉糰。
2. 熟粉糰放入大窩內，加入糯米粉 310 克、溫水 250 克及生油 1.5 湯匙，拌勻揉成糯米糰，備用。

湯圓做法

1. 湯料用慢火煮約 15 分鐘，備用。
2. 冬菇用清水浸軟，洗淨、蒸熟（做法參考 p.17）。
3. 貢菜用清水浸泡約 30 分鐘至味道略淡，備用。
4. 豬頸肉放入潮州滷水汁內（滷水汁做法參考 p.58「潮州滷水鵝」），慢火煮 30 分鐘，盛起。
5. 冬菇、貢菜及滷水豬頸肉切成幼粒，備用。
6. 白芝麻放入白鑊內，用慢火炒至金黃色。
7. 燒熱鑊下生油，用慢火爆香紅葱頭，再放入其他餡料，拌炒後潷酒，加入老抽及蠔油炒勻，勾芡，熄火，加入熟芝麻拌勻成湯圓餡料。
8. 取湯圓皮約 30 克，搓成圓球形，略壓扁，包入餡料 1 湯匙包實，搓成湯圓備用。
9. 煮熱湯水，放入湯圓翻滾，轉中火煮約 5 分鐘至湯圓浮起，盛起食用。

燒味·節慶篇

燒味・小食篇

Snack and appetizers

燒味，包含燒、臘、滷、燻，

用這些烹調手法製成富有特色的派對小食，

滷水冷盤應有盡有，

掀起派對的高潮！

水晶肴肉

Jellied pork cold cut

材料

鹹豬手	2	隻
新鮮雞腳	5	隻
淨豬皮	300	克
薑片	30	克
玫瑰露酒	3	湯匙
麻油	1	湯匙
魚膠粉（後下）	30	克

滷水香料

甘草	2	片
八角	1	粒
香葉	4	片
草果	2	粒
桂皮	1	克
沙薑	5	片
丁香	8	粒
小茴香	1	湯匙
薑片	30	克
葱	30	克

滷水調味料

清水	10	杯
砂糖	20	克
幼鹽	40	克
雞粉	10	克
花雕酒	20	克
白胡椒粉	半	茶匙

蘸汁

鎮江香醋	3	湯匙
薑絲	10	克

燒味 · 小食篇

掌握重點技巧

用方盤壓着肴肉表面時，建議加上重物增加壓力，緊壓肴肉，令肴肉凝固結實，以免肉質鬆散。

做法

1. 滷水香料用煲湯魚袋包好，放入滷水調味料內用慢火煲約半小時成滷水汁。

2. 鹹豬手及豬皮燒去細毛；雞腳剪去趾甲，將以上材料洗淨，放入滾水內，加入薑片 30 克及玫瑰露酒 3 湯匙煮 5 分鐘，取出洗淨。

3. 鹹豬手、豬皮及雞腳放入滷水汁內煮滾，轉慢火煲 3 小時。取出豬手去皮、起肉、去骨，豬肉切成粗粒狀，滷水汁留用。

4. 魚膠粉用清水 150 毫升調溶（或座於熱水至溶），加入滷水汁內拌勻，放入豬皮及肉粒浸 5 分鐘，取出。

5. 在方盤底塗抹少許麻油，鋪上一塊豬皮，再疊上一層豬手肉，加入滷水汁至剛浸過豬手肉，最後放上豬皮蓋好。

6. 放上另一個方盤壓平表面，放入雪櫃冷藏 4 小時至凝固及定型，切件蘸汁享用。

燒味・小食篇

丁香鴨舌 Duck tongues in spiced marinade

材料

急凍鴨舌	400 克

料頭

蒜粒	20 克
乾葱粒	20 克
薑粒	20 克
生油	2 湯匙
花雕酒	1 湯匙

滷水料

清水	760 克
生抽	130 克
蠔油	150 克
老抽	30 克
幼鹽	5 克
片糖	100 克
雞粉	30 克
八角	2 粒
陳皮	1 個
甘草片	5 片
川椒	5 克
小茴香	10 克
丁香	10 克

做法

1. 起鑊下油爆香料頭,加入滷水料,用中火煲滾,轉慢火煮 10 分鐘,熄火,加蓋浸焗至散出香味,待涼,隔去渣滓備用。

2. 鴨舌解凍後,放入滾水內,待翻滾後熄火,浸 10 分鐘,取出鴨舌,用清水沖洗至涼爽,瀝乾水分。

3. 鴨舌放入步驟1的滷水汁內,冷藏浸泡 2 小時即可。

燒味・小食篇

醬烤鳳翼扎

Stuffed chicken wings wrapped in pork belly

材料

急凍雞全翼	6 隻
原條去皮五花腩	約 300 克
甘筍	60 克
西芹	60 克
京葱	60 克
芫茜	30 克
百花蜜	300 克

醃料

砂糖	12 湯匙
幼鹽	2 湯匙
雞粉	1 湯匙
海鮮醬	4 湯匙
蠔油	4 湯匙
芝麻醬	1 湯匙
生抽	半湯匙
老抽	半湯匙
花雕酒	3 湯匙
蒜茸	2 湯匙
乾葱茸	1 湯匙
雞蛋	2 隻

 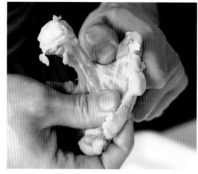 

做法

1. 將去皮五花腩肉平放盤內，放入冰箱急凍至凝固，取出，切成 6 片長薄片。

2. 雞翼保留翼尖骨，用剪刀褪出肉、去掉大骨，洗淨、瀝乾，與醃料拌勻醃 3 分鐘，取出備用。

3. 芫茜去葉；甘筍去皮；西芹刨去根；京葱洗淨，全部切成約 5 吋長粗條。

4. 腩肉片放入醃料內拌勻，備用。

5. 每隻雞翼內釀入芫茜梗 5 條、西芹 10 條、甘筍 10 條及京葱絲，將腩肉片由翼尖位斜捲包緊釀料，最後在腩肉片尾部穿上牙籤固定收口。

6. 預熱焗爐，雞翼扎平放在烘架上，放入用錫紙包好的底盤，用 240℃ 烘 5 分鐘，反轉另一面再烘 5 分鐘，調低溫度至 125℃ 烘 10 分鐘，取出。

7. 用百花蜜塗勻雞翼，放進焗爐用 240℃ 烘 3 分鐘，再轉另一面烘 3 分鐘，調至 125℃ 烘 10 分鐘，取出抹一遍百花蜜即成。

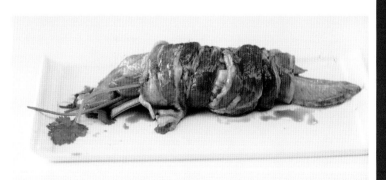

掌握重點技巧

1. 去骨雞翼及薄腩片較易入味，因此醃製時間不要太久，以免過鹹。

2. 包扎雞翼時，若腩片長度不足以完全包裹雞翼，可多加一塊腩片緊接，至完全包好雞翼。

燒味・小食篇

137

滷香燻魚

Smoked fish in sweet marinade

材料

鮎魚腩	1 斤

醃料 A

清水	450 克
薑片	40 克
白胡椒粉	1/4 茶匙
幼鹽	15 克
砂糖	10 克
雞粉	5 克
魚露	10 克

醃料 B

蒜肉	10 克
乾葱肉	30 克
米酒	20 克

滷香汁

生油	2 湯匙
薑片	20 克
葱	20 克
乾葱肉	20 克
八角	3 粒
花椒	1 茶匙
米酒（後下）	1 湯匙

調味料

清水	600 克
金獅糖漿	50 克
照燒醃醬（包裝）	55 克
海鮮醬	60 克
蠔油	20 克
幼鹽	1 茶匙
雞粉	1 茶匙
老抽	1 茶匙
潮汕米醋	1 茶匙
香葉	5 片
桂皮	1 克
沙薑	6 粒
白胡椒粉	1/4 茶匙
辣椒油	1 茶匙

掌握重點技巧

1. 切魚件時寬度不可太薄，否則容易炸乾及散爛。

2. 再燒熱油翻炸魚件，可去除魚腥味，而且令魚肉容易吸收滷香汁之味道。

燒味·小食篇

139

滷香汁做法

燒熱鑊，下生油爆香滷香汁材料，加入調味料煮滾，轉慢火熬半小時，待涼，隔去材料備用。

做法

1. 鯇魚腩洗淨，切件，每件約 2cm 寬度。
2. 醃料 A 用慢火煮 10 分鐘，待冷卻後，加入醃料 B 浸泡 40 分鐘，放入魚件醃 25 分鐘，盛起放於笪箕，瀝乾水分，用廚房紙吸乾魚肉表面水分。
3. 燒熱油至 8 成溫，放入魚件炸約 1 分鐘，轉中慢火炸至金黃色，盛起。再加熱油溫，放入魚件翻炸至表面酥脆，瀝去油分。
4. 將炸好的魚件放入滷香汁內，用慢火煮 5 分鐘，取出魚件待涼。
5. 魚件放入碟內排好，倒入已涼的滷香汁，浸泡 2 小時至香味滲透魚肉即可。

煙燻素鵝

Smoked vegetarian beancurd skin roll

掌握重點技巧

1. 包扎素鵝卷時，餡料必須平均分佈，而且將腐皮包緊餡料，以免鬆散令餡料溢出。

2. 用作煙燻的鑊必須先鋪上錫紙，才放上煙燻料，否則容易令鑊焦黑。

3. 煙燻料內加上麵粉，令氣味更香；待砂糖燒至微溶時，才放入腐皮卷煙燻。

材料

大腐皮	2 張
麻 油	2 茶匙
錫 紙	1 張

餡料

生 油	2 湯匙
冬菇絲	120 克
（做法參考 p.17）	
西 芹	100 克
甘 筍	100 克
木 耳	1 朵

調味料

開 水	120 克
砂 糖	1 茶匙
幼 鹽	半茶匙
蠔 油	2 茶匙
五香粉	半茶匙
生 粉	2 茶匙

煙燻料

茶 葉	20 克
麵 粉	20 克
砂 糖	30 克

做法

1. 木耳用清水浸軟，切幼絲；西芹及甘筍洗淨，去皮，切成幼絲。

2. 調味料拌勻，預留約 40 克作掃面之用。

3. 燒熱油，下餡料炒香，加入調味料 90 克炒勻至水分收乾，盛起，待涼。

4. 腐皮剪去硬邊，在腐皮兩面掃上已預留之調味料，對摺成半圓形，放入炒香的餡料，包緊捲成約 2 吋寬的長條形。

5. 素鵝卷掃上麻油，放入蒸籠蒸 5 分鐘，取出冷卻。

6. 燒熱鑊，下油 2 湯匙，放入素鵝用慢火煎至金黃香脆，盛起。

7. 鑊內鋪上錫紙，下煙燻料平均分佈，放上格網及排好素鵝卷，用中火燒熱至燻料出煙，加蓋，轉慢火燻 2 分鐘，熄火焗 2 分鐘，取出，掃抹麻油，切件上碟。

燒味・小食篇

143

鵝肝燒肉腸

Grilled homemade pork
and foie gras sausage

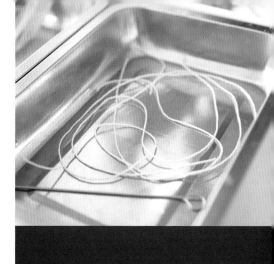

肉腸材料

半肥瘦梅肉	300 克
熟冬菇	6 朵
（做法參考 p.17）	
馬蹄肉	80 克
鵝肝	80 克
芫茜粒	1 湯匙
生粉	2 茶匙
雞蛋	1 隻

材料及工具

人造腸衣	150cm
線繩	60cm
漏勺	1 個

醃料

砂糖	5 茶匙
幼鹽	1 茶匙
雞粉	半茶匙
海鮮醬	2 茶匙
蠔油	2 茶匙
芝麻醬	1 茶匙
生抽	1 茶匙
老抽	半茶匙
蒜茸	1 茶匙
乾葱茸	3 茶匙
玫瑰露酒	1 茶匙

蘸料

蘇梅醬	2 湯匙
糖桂花	2 湯匙

掌握重點技巧

1. 將餡料灌入腸衣時，若腸衣內積存空氣令餡料難以滑下，可用小針在打結位稍刺二至三次，空氣排出後，餡料可順利向下滑至填滿腸衣。

2. 人造腸衣在西環售賣自製臘腸的店舖有售。

燒味・小食篇

即時睇片

做法

1. 人造腸衣用清水浸軟，剪成三段（每段約 50cm）；線繩剪成三段，每段約 20cm，用作打結封口之用。

2. 醃料預先調勻，備用。

3. 熟冬菇、馬蹄肉、梅肉及鵝肝一同切成幼粒，加入芫茜、生粉及雞蛋混合，下醃料 2 湯匙拌勻醃 5 分鐘，即成肉腸餡料。

4. 在腸衣的其中一端打結，另一端插入漏勺，將肉腸餡料灌入腸衣內，加滿後用線繩打結，用醃料在肉腸表面塗勻醃 5 分鐘。

5. 預熱焗爐 150℃，肉腸平放於已包錫紙的烘盆上，用幼針在肉腸近兩端的打結位戳入鬆針，使空氣排出。

6. 肉腸放入焗爐，用 150℃ 烘 15 分鐘後，爐溫調低至 100℃，反轉肉腸再烘 15 分鐘取出，伴蘸料享用。

Lettuce wrap with
beef shin and
flying fish roe

材料

滷水牛䐉	8	片
即食紫菜片	2	張
即食酸子薑	40	克
原塊腐皮	4	張
蛋白	2	湯匙
飛魚子	8	湯匙
千島沙律醬	8	湯匙
羅馬生菜	12	片
大青瓜	1	條

湯味料

清水	2	杯
砂糖	2	茶匙
雞粉	1	茶匙
麻油	1	湯匙
鮮醬油	半	茶匙
鹽	1	茶匙

燒味・小食篇

149

掌握重點技巧

1. 滷水牛腱的製法可參考第三十六頁「紅油牛腱」，或自購原件滷牛腱切薄片亦可。

2. 牛腱滷製後，必需冷藏才可切成薄片，否則切片時容易弄散腱肉。

做法

1. 湯味料煲滾後，熄火備用。

2. 滷水牛腱切薄片，每片約 1.5 吋 x 3 吋；即食紫菜剪成每片約 6 吋 x 6 吋；即食酸子薑切粗絲。

3. 羅馬生菜修剪成樹葉形；青瓜橫切成圓厚片，中間挖空成圓圈形狀。

4. 取一塊腐皮剪成 4 小張（每張 6 吋 x 6 吋正方形），浸於湯味料內至鬆軟及呈鮮黃色，備用。

5. 另一塊原張腐皮放入湯味料內浸軟，取出張開平放，對摺成半圓形狀。

6. 取兩小片泡鬆軟的腐皮，平放在半圓形近底橫線的中間位置，排上紫菜片及餘下的兩小張腐皮，再放上牛腱片平排，將兩側腐皮反摺中間，最後放上子薑絲，用力捲起，塗上蛋白封口。

7. 腐皮卷用保鮮紙包好，用針在保鮮紙表面均勻地戳上小孔，隔水蒸 30 分鐘，待涼後放入雪櫃貯存。

8. 食用時將金牛卷斜切成六小件，塗上千島沙律醬及灑勻飛魚子，放於青瓜環上，伴生菜葉享用。

椰香白雲鳳爪

Chicken feet in coconut milk marinade

材料

雞腳	600 克（約 15 隻）
椰皇	1 個
薑	20 克
葱	20 克

滷水料

清水	1 公斤
小茴香	5 克
沙薑粒	5 克
八角	2 粒
香葉	5 克
薑片	20 克

滷水調味料

椰奶（小罐裝）	165 毫升
椰皇水	1 個
幼鹽	50 克
冰糖	15 克
雞粉	20 克
花雕酒	30 克

蘸料

泰國雞醬	適量

滷水汁做法

1. 將椰皇水盛起，取出椰肉切片，備用。
2. 滷水料煲滾後，轉慢火煮 15 分鐘，加入滷水調味料煮至冰糖完全溶解，熄火待冷，隔去滷水料備用。

做法

1. 雞腳洗淨，剪去趾甲，與薑及葱各 20 克放入水內，翻滾後轉文火煮 10 分鐘，熄火，加蓋焗 35 分鐘。
2. 取出雞腳用冰開水浸泡至涼爽，將雞腳切開兩邊，備用。
3. 雞腳及椰肉放入滷水汁內，貯存雪櫃泡浸約 4 小時或以上，享用時可蘸泰國雞醬。

掌握重點技巧

1. 雞腳不可用大火直接煲腍，否則令雞腳外皮爆裂或過腍。
2. 雞腳軟腍後，必須盡快降溫，浸冰水使雞腳外皮收縮，帶爽脆口感。
3. 先將雞腳切開兩邊，再浸入滷水汁內，除了更入味外，進食時也更方便。

燒味・小食篇

Beef cheek in Sichuan spiced marinade

Ingredients
600 g beef cheek

Spices
4 tbsp cooking oil
25 g sliced ginger
20 g shallot (finely chopped)
20 g grated garlic
1 red chilli
2 tbsp Sichuan peppercorns
1 clove star anise
1 tbsp spicy bean sauce
1 tbsp Sichuan chilli sauce

Marinade
300 g water
150 g light soy sauce
150 g brown sugar slab
1/2 tsp table salt
1 tsp five-spice powder
1/2 tbsp chicken bouillon powder
40 g Chinese celery
15 g spring onion
1 tbsp Sichuan pepper oil (added at last)
1 tbsp Shaoxing wine (added at last)

Method

1. Rinse the beef cheek. Blanch in boiling water. Bring to the boil again over high heat. Then turn to low heat and cook for 1 hour. Soak it in ice water to chill instantly. Drain and set aside.

2. To make the marinade, heat a pot and add oil. Fry the spices until fragrant. Add the marinade ingredients. Bring to the boil. Put in the beef cheek from step 1. Turn to low heat and simmer for 15 minutes. Then turn off the heat and cover the lid. Leave it in the marinade for 15 minutes.

3. Take the beef out and leave it to cool completely. Refrigerate. Store the marinade for later use. Before serving, slice the beef and drizzle with warm marinade.

Tips

· You have to reserve beef cheek from the butcher. If they run out of beef cheek, use beef shin instead.

· Beef cheek has fine and dense muscle fibres, which account for its chewy and springy texture.

· I personally prefer Sichuan chilli sauce from Lao Gan Ma brand. I love its strong rich aroma and the numbing sensation it gives.

Pork trotter in white pepper stock

Ingredients
12 pieces frozen pork trotters
10 g ginger
10 g spring onion

Spiced marinade
1.5 kg water
1 clove star anise
40 g Chinese celery
20 g ginger
20 g spring onion
3 bay leaves
15 g white peppercorns
40 g table salt
20 g chicken bouillon powder

White pepper stock
1.2 kg water
25 g white peppercorns
1 clove star anise
2 bay leaves
10 g table salt
10 g chicken bouillon powder

Method
1. Fry the white peppercorns in the spiced marinade and the white pepper stock separately in a dry wok until fragrant. Crush them with the flat side of a knife slightly.

2. Boil the spiced marinade. Turn to low heat and cook for 10 minutes. Turn off the heat and leave it to cool.

3. In another pot, boil the white pepper stock. (Put the crushed white peppercorns in a muslin bag and tie well before putting them in the stock.) Turn to low heat and cook for 10 minutes. Turn off the heat.

4. Thaw the pork trotters. Burn off the hair with a kitchen torch. Clip the nails. Rinse well.

5. Boil a pot of water and put in the pork trotters, ginger and spring onion. Bring to the boil again. Turn to low heat and simmer for 1 hour 30 minutes until the pork is tender. Remove from heat and rinse in cold water until the skin no longer feels sticky. Drain.

6. Strain the marinade. Put the pork trotters in and keep in the fridge. Leave them for 4 hours.

7. Before serving, bring the white pepper stock to the boil. Put the pork trotters in. Cook over low heat for 3 minutes. Transfer onto a serving plate. Serve.

Tips
· I prefer white peppercorns from Kam Fook brand because of its powerful kick and aroma.
· After you put the blanched pork trotters in the marinade, make sure you keep them in the fridge. Just heat it up again in the white pepper stock before serving.
· I suggest chopping the pork trotter into 8 small pieces first if the whole pork trotter is used.

Dongpo pork pagoda

Method

1. Blanch pork belly in boiling water for 5 minutes. Drain. Rinse in cold water. Set aside.

2. To make the marinade, heat 2 tbsp of oil in a wok. Stir fry the spices until fragrant. Add marinade ingredients and bring to the boil. Put a bamboo mat on the bottom of the pot to prevent the pork from burning. Put in the blanched pork belly. Bring to the boil and turn to low heat. Cover the lid and cook for 2 hours until pork turns tender. Remove the pork and leave it to cool. Strain the marinade and save for later use.

3. Put the pork on a tray with the skin down. Leave in a quick freezer until frozen. Slice into 4mm thick slices. Set aside.

4. Cut the Mei Cai into small florets. Soak in water for 1 hour. Rinse off the sand. Drain and finely chop it.

5. To make the glaze, heat a wok. Add 2 tbsp of oil. Stir fry shallot until fragrant. Put in Mei Cai and stir well. Stir in seasoning for thickening glaze. Cover the lid and cook over low heat for 15 minutes. Set aside. Keep the glaze for later use.

6. Stack the sliced pork at an angle and fold one slice into another to build a pagoda. Put into a pagoda mould. Stuff the pagoda with Mei Cai. Press to compact the Mei Cai. Steam the pork pagoda in the mould for 30 minutes (the pagoda point is downward). Remove the pork pagoda and turn it out of the mould. Transfer onto a serving plate. Bring the leftover glaze from step 5 to the boil and stir in thickening glaze. Drizzle over the pork pagoda. Serve.

Ingredients
600 g boneless pork belly
300 g sweet Mei Cai
2 tbsp cooking oil
50 g sliced shallot

Spices
20 g ginger
20 g spring onion
20 g garlic cloves
20 g shallots
3 cloves star anise

Marinade
750 g water
45 g brown sugar slab
230 g premium gold soy sauce
230 g Shaoxing wine
1 tsp table salt
2 tsp chicken bouillon powder
1 tsp five-spice powder
10 cloves

Seasoning for thickening glaze
10 tbsp Shaoxing wine
500 g marinade
300 g water

Thickening glaze
1 tbsp potato starch
1 tbsp water

Tips

· You have to stew the pork belly until tender and then freeze it so as to slice it easily and thinly.

· I prefer Lee Kum Kee's premium gold soy sauce because it tastes quite different from regular light soy. It's less salty but with a strong soybean taste.

· You can buy the pagoda mould from the stores specializing in cooking utensil in Shanghai Street. The mould model is H28 for cake.

Home-style recipes

Deep fried squabs with fermented beancurd sauce

Ingredients
2 squabs (about 480 g each)

Spice stock
3 litres water
40 g sliced ginger
40 g lemongrass
5 cloves star anise
20 bay leaves
2 tbsp cloves
10 g cassia bark
10 slices liquorice
5 g white peppercorns
20 slices sand ginger
4 tbsp cumin seeds

Fermented beancurd marinade
340 g rock sugar
230 g table salt
230 g white fermented beancurd
40 g chicken bouillon powder
40 g Shaoxing wine
40 g Chinese rose wine

Basting sauce
50 g Basting sauce A (refer to p.13 for method)

Dipping sauce (mixed well)
3 cubes white fermented beancurd
2 tbsp Shaoxing wine
1 tbsp sugar
1 tsp sesame oil

Method
1. Bring spice stock to the boil. Turn to low heat and cook for 1 hour. Add fermented beancurd marinade ingredients. Cook until rock sugar dissolves. Turn off the heat.

2. Cut off the feet of the squabs. Trim off the fat around the tail. Remove the lungs and windpipes. Rinse well. Put them into the boiled marinade from step 1. Turn on high heat to bring to the boil again. Turn to low heat. Cook for 5 minutes. Turn off the heat and cover the lid. Leave them in for 25 minutes.

3. Take the squabs out. Rinse off the fat on the skin with hot water. Brush basting sauce over the skin. Dry with a hair dryer or over low heat in an oven.

4. Heat oil in a wok. Deep fry the squabs until golden and crispy. Serve with dipping sauce.

Tips
· If you dry the squabs in an oven, make sure you don't overheat them. Otherwise, the maltose will melt and burn the skin. Just keep the oven at 100°C and leave the door ajar.

· After you blow dry or oven-dry the basting sauce, it should not feel sticky when you touch it.

· The dipping sauce heightens the flavours of the squabs. It is an innovative way to serve fried squabs.

Barbecue pork on hot plate with osmanthus molasses

Ingredients

1 kg pork shoulder butt
1 onion
1 bottle multi-floral honey (used for barbecue)
8 tbsp potato starch

Marinade

5 tbsp sugar
1 tbsp table salt
1/2 tbsp chicken bouillon powder
2 tbsp Hoi Sin sauce
2 tbsp oyster sauce
1/2 tbsp sesame paste
1/2 tsp light soy sauce
1/2 tsp dark soy sauce
1/2 tbsp finely chopped shallot
1 tsp grated garlic
1/2 tbsp Chinese rose wine
1 egg

Osmanthus molasses

60 g water
40 g brown sugar slab
20 g light soy sauce
120 g multi-floral honey
10 g salt
1 tsp dried osmanthus

Method

1. Cut the pork into long strips about 11 inches long, 2.5 inches wide and 0.8 inch thick. Rinse in water for 30 minutes. Drain. Add 4 tbsp of potato starch and rub it evenly on the pork. Refrigerate for 4 hours.

2. Rinse off the potato starch and the blood from the pork. Add 4 tbsp of potato starch again. Rub evenly. In a mixing bowl, mix the marinade well. Put the pork in and toss to coat it in marinade. Leave it for 40 minutes.

3. To make the osmanthus molasses, boil the ingredients. Set aside.

4. Preheat an oven to 240°C. Arrange the pork flat on a grilling rack. Put the rack over a baking tray wrapped in aluminium foil. Bake the pork for 10 minutes. Flip the pork upside down. Bake for 10 more minutes until lightly charred. Turn to low heat at 100°C and bake for 30 more minutes. Take the pork out. Brush multi-floral honey on it.

5. Turn the oven back to high heat at 240°C. Put the pork in and grill until the molasses bubbles. Turn to low heat at 100°C again. Bake for 30 minutes. Take the pork out and brush on multi-floral honey again.

6. Heat a cast iron platter on stove. Slice the barbecue pork from step 5 thinly at an angle. Coarsely dice the onion.

7. On a hot iron platter, put on diced onion and arrange sliced barbecue pork neatly. Drizzle with osmanthus molasses at last. Serve.

Tips

· When you rub potato starch into the pork, make sure you add a little water. That helps form a thin paste on the surface of the pork, to facilitate the osmotic action.

· Marinating the pork with potato starch for 4 hours helps remove any impurities and blood on the pork. It also tenderizes the muscle to some extent.

· Make sure the oven is preheated to desired temperature before you put the pork in. Otherwise, the meat juices and marinade cannot be sealed in within 10 minutes. The pork may dry out.

· At high temperature (240°C), the oven scars the surface of the pork to seal in the juices. At low temperature (100°C), the oven cooks through the meat slowly. The pork will turn out more succulent and tasty that way.

Spicy soy-marinated beef shin

Ingredients
1 frozen beef shin (about 500 g)

Spices
2 tbsp cooking oil
20 g shallot
10 g garlic cloves
20 g sliced ginger
25 g Chinese celery
20 g spicy bean sauce
2 bird's eye chillies (diced)

Marinade
1.2 kg water
160 g light soy sauce
40 g Maggi's seasoning
60 g Shaoxing wine
1 tbsp dark soy sauce
1 tbsp Sa Cha sauce
20 g oyster sauce
10 g table salt
20 g chicken bouillon powder
1 tbsp five-spice powder
1 tbsp chilli powder
160 g brown sugar slab

Method
1. Thaw the beef shin. Blanch in boiling water for 5 minutes. Drain and rinse in cold water for 5 minutes. Pierce the surface of the beef with skin pricker (or a fork, a bamboo skewer) evenly for the marinade to infuse.

2. In a pot, stir fry the spices until fragrant. Add the marinade and bring to the boil. Put in the beef shin and bring to the boil. Turn to low heat and cover the lid. Cook for 1 hour 30 minutes until the beef is tender. Remove the beef. Leave it to cool.

3. Strain the marinade and leave it to cool. Put in the cooled beef shin. Keep in the fridge.

4. Before serving, slice the beef shin thinly. Drizzle with the marinade. Serve.

Tips
· When you make the marinade, make sure you stir until sugar dissolves. Otherwise, it may burn and stick to the bottom.
· When you cook the beef shin in the marinade over low heat, you must cover the lid. The beef will turn tender more easily and both the aroma and moisture are preserved.

Duck feet rolls in molasses

Method

1. Blanch the deep fried duck feet in boiling water. Bring to the boil again and turn to low heat. Simmer for 30 minutes. Drain. Soak in cold water until cooled.

2. Trim off the fat on the chicken intestines. Add 1 tbsp of potato starch and 1/2 tbsp of salt. Rub well. Rinse. Put the chicken intestines in recently boiled water until they shrink. Rinse in cold water until cool. Drain.

3. Cut the Zha Cai into 2.5-inch long strips about the thickness of a chopstick. Rinse in water to remove saltiness. Set aside.

4. Cut the pork tripe and barbecue pork into 2.5-inch long strips. Cut the chicken livers into halves. In a bowl, put in the chicken liver and duck feet. Add enough marinade to cover. In another bowl, put in the chicken intestines and add enough marinade to cover. (It's hard to retrieve the chicken intestines if marinated with other ingredients.)

5. Lay flat a duck feet. Put one strip each of Zha Cai, pork tripe, barbecue pork, and chicken liver over the duck feet. Wind chicken intestines around the ingredients into a bundle. Secure both ends with toothpicks. Brush Char Siu sauce over it. Leave it for 5 minutes.

6. Preheat an oven to 240°C. Bake each side of the duck feet rolls for 10 minutes. Brush on maltose. Bake each side for 5 more minutes at 200°C. Brush on one more coat of maltose. Serve.

Ingredients

6 deep fried duck feet
60 g spice-marinated pork tripe
60 g barbecue pork
60 g Zha Cai (Sichuan spicy mustard tuber)
3 chicken livers
180 g chicken intestines
40 g maltose
2 tbsp Char Siu sauce
12 bamboo toothpicks

Marinade

9 tbsp sugar
3 tsp table salt
1.5 tbsp chicken bouillon powder
1.5 tbsp light soy sauce
1.5 tbsp red fermented beancurd
1.5 tbsp pickling brine from
red fermented beancurd
3 tbsp Hoi Sin sauce
2 tbsp sesame paste
2 tbsp Chinese rose wine
1.5 tbsp grated ginger
3 tbsp finely chopped shallot
3 tbsp grated garlic
1.5 tsp five-spice powder
2 tsp potato starch
2 eggs
2 tbsp Shaoxing wine

Tips

· You may use goose intestines or even thinly sliced pork belly instead of chicken intestines.

· The chicken intestines come with much fat clinging on them. Make sure you trim it off carefully. Do not soak them in hot water for too long. Otherwise, it shrinks too much and becomes rubbery.

· When you wind the chicken intestines around the filling, do it so that the smoother side of the intestines face the inside of the roll. With the darker and coarser side on the outside, it picks up the colour and flavour of Char Siu sauce more easily.

· If one piece of chicken intestine isn't enough to wind around the roll from end to end, you may join it with another intestine.

· After you wind the chicken intestines around the roll, secure them by inserting two toothpicks to make a cross. Try to avoid piercing through the bones in the duck feet.

· I prefer Char Siu sauce from Tung Chun brand. It adds background seasoning to the outside of the duck feet rolls and gives it a beautiful caramel colour after baked.

Home-style recipes

Soy marinated chicken in casserole

Ingredients

1 dressed chicken (about 1.35 kg)
160 g Chinese rose wine
160 g Shaoxing wine
40 g dark soy sauce

Marinade

3 litres water
3 litres light soy sauce
20 g red yeast rice
1 whole dried tangerine peel
10 g liquorice
10 g bay leaves
10 g cassia bark
5 g cloves
20 g diced sand ginger
10 g cumin seeds
5 cloves star anise

Seasoning for marinade

2 kg rock sugar
350 g table salt
100 g chicken bouillon powder

Dressing sauce

80 g hot water
30 g light soy sauce
25 g oyster sauce
60 g sugar
1 tsp dark soy sauce
1 tsp chicken bouillon powder
1/4 tsp table salt

Aromatics

2 tsp cooking oil
5 shallots (halved)
5 garlic cloves (halved)
20 g spring onion (cut into short lengths)
5 big slices ginger
1 tbsp Shaoxing wine

Method

1. To make the marinade, boil the ingredients. Turn to low heat and cook for 30 minutes until fragrant. Turn off the heat. Add seasoning for marinade. Stir until sugar dissolves.

2. To make the dressing sauce, boil all ingredients. Turn off the heat and set aside.

3. Remove the innards and solid fat inside the chicken. Rinse well. Boil the marinade from step 1. Put in the chicken and bring to the boil again. Turn to low heat and cook for 5 minutes. Turn off the heat and leave the chicken in the hot marinade for 35 minutes until cooked through. Take the chicken out. Chop into pieces and set aside.

4. Heat a casserole pot. Add 2 tsp of oil. Stir fry the aromatics until fragrant. Sprinkle with Shaoxing wine. Put in the chicken. Drizzle with the dressing sauce from step 2. Cover the lid and bring to the boil. Serve in the casserole.

Tips

· It's advisable to cook the marinade in a tall but narrow stainless steel pot. It will create some depth for the marinade to cover the chicken. That helps the chicken to cook through and the flavours to infuse.

· You may store the spiced marinade for later use. Just boil it again and leave it to cool. Keep it in a fridge.

· Red yeast rice is available from stores specializing in spices.

Method
Flat bread

1. Mix hot water with salt and cooking oil. Pour in cake flour and stir quickly. Knead into smooth dough. Cover with damp cloth and let cool.

2. Roll the dough into a cylinder. Cut into 24 segments. Then roll each small piece of dough out into a flat round patty. Fry in a dry pan over low heat to fix its shape. Before serving, steam the flat bread for 2 minutes.

Peking duck

1. Trim off the solid fat on the insides of the duck. Remove lungs, windpipe and other innards. Rinse and drain. Mix the marinade and rub it on the inside of the duck. Secure the tail with a metal skewer.

2. Boil water. Dip the duck in the boiling water while scooping up hot water with a ladle to pour over its skin all over. This shrinks the skin and changes its colour. Rinse in cold water until the duck skin is cool and no longer sticky. Drain. Hang the duck by securing two hooks on the inner side of the wings. Drape the duck head on the back side. Brush basting sauce B over the skin. Hang the duck until dry.

3. Preheat an oven to 150 °C. Put the duck on a baking tray with the breast side up. Remove the metal skewer. Bake the duck for 50 minutes until lightly browned. (You should check the duck after baking for 30 minutes. If it looks browned, cover it with aluminium foil.) Turn the heat down to 100 °C and bake for 30 minutes until the duck is done. Take it out of the oven and hang it. Drain any liquid inside the duck.

4. Heat oil in a wok and put in the duck to fry it until nicely browned. Then slice the skin with some meat off it.

5. Steam the flat bread. Before eating, each guest wrap his own the duck pieces in the flat bread with Peking scallion, cucumber strips and Hoi Sin sauce.

Peking duck

Ingredients
1 duck (about 2.4 kg)
200 g cucumber
150 g Peking scallion
4 tbsp unforgettable sauce
(refer to p.15 for method)
24 sheets homemade flat bread

Basting sauce
150 g Basting sauce B
(refer to p. 14 for method)

Marinade
1.5 tbsp sugar
1 tbsp table salt
1 tsp chicken bouillon powder
1/4 tsp five-spice powder
3 cloves star anise
1 whole dried tangerine peel
2 tbsp Shaoxing wine

Dough for flat bread
250 g cake flour
200 g hot water at 80 °C
1/4 tsp salt
1/2 tsp cooking oil

Tips

· After putting the duck in hot oil, there may be bubble puffing up on the skin. If that happens, use a metal skewer to pierce through the skin and release the bubble. Otherwise, the duck may have bumpy skin and doesn't look nice.

· If you want to save trouble of making your own flat bread, you can buy readymade ones for Peking duck from stores specializing in noodles.

Banquet recipes

163

Grilled lamb chops in Portuguese sauce

Ingredients

6 lamb chops (about 100 g each)
12 tbsp potato starch
8 tbsp water
12 cherry tomatoes
1 purple onion (finely shredded)

Marinade

10 tsp sugar
2 tsp table salt
2 tsp chicken bouillon powder
1.5 tsp ground cumin
1/2 tsp five-spice powder
1/2 tsp ground black pepper
2 tsp Hoi Sin sauce
2 tsp oyster sauce
1 tsp sesame paste
2 tsp gold premium soy sauce
2 tbsp rice wine
2 tsp finely chopped shallot
2 tsp grated garlic
2 tsp grated ginger

Portuguese sauce

1 tbsp curry paste
4 tbsp coconut milk
1 tbsp ketchup
1 tsp sugar
1 tsp salt
1 tsp chicken bouillon powder
20 g butter
2 cups chicken stock
3 tbsp potato starch

Method

1. Mix the marinade ingredients together one day ahead. Keep in the fridge.

2. Rinse the lamb chops. Add 6 tbsp of potato starch and 4 tbsp of water. Mix well. It should look like a thin paste. Leave them for 30 minutes. Rinse well.

3. Add 6 tbsp of potato starch and 4 tbsp of water to the lamb chops again. Mix well into a thin paste. Then stir in the marinade from step 1. Leave them for 30 minutes.

4. Heat oil in an oven-safe skillet. Fry the lamb chops over medium heat until both sides golden. Set aside.

5. In the same skillet, put in half of the shredded purple onion. Arrange lamb chops on top with cherry tomatoes.

6. In another pan, melt butter and stir fry the remaining purple onion until fragrant. Add the Portuguese sauce ingredients. Stir and bring to the boil. Thicken with potato starch slurry at last. Bring the boil and pour over the lamb chops.

7. Preheat an oven up to 200 °C. Transfer to skillet into the oven. Bake until lightly browned. Serve.

Tips

· Mixing the marinade one day ahead helps the flavours to mingle and enhance each other.

· In step 2, marinating the lamb chop with a thin potato starch slurry serves two purposes. First it cleanse the lamb chop by attracting impurities and blood. Second, it form a thin membrane around the lamb chop, so as to keep it moist and seal in juices.

· Those who prefer their lamb chop to taste gamey should shop for those from Mongolia. They tend to have a stronger flavour.

Fried chicken in red tarocurd sauce

Method

1. Rinse the chicken. Put it in a pot of boiling water to shrink the skin. Rinse with cold water until the skin is no longer sticky. Set aside.

2. Put all marinade ingredients into a blender. Blend until fine. Pour into a tray. Put the chicken in and spread evenly on the skin and insides. Leave it for 1 hour.

3. Cut the chicken along the breast bone. Pull apart and flatten it into a Pipa shape by inserting metal skewers or bamboo chopsticks on the inside. Rinse the skin with water. Drain. Brush on the basting sauce. Leave it to air dry or dry it in an oven.

4. Preheat an oven to 150 °C. Put in the chicken with the skin side up. Bake for 30 minutes. Turn to 100 °C and bake for 30 more minutes. Hang the chicken up. Prick the skin with a needle.

5. Heat oil in a wok. Hang the chicken over the wok. Scoop oil in a ladle and pour it over the chicken the crisp up the skin. Chop the chicken up and serve with the dipping sauce that has been boiled.

Ingredients
1 dressed chicken (about 1.35 kg)

Basting sauce
120 g basting sauce A (refer to p.13 for method)

Marinade
680 g red tarocurd
680 g light soy sauce
80 g white fermented beancurd
80 g Shaoxing wine
80 g shallot
160 g garlic cloves
40 g coriander
80 g spring onion
200 g table salt
120 g chicken bouillon powder
80 g ginger
40 g galangal
100 g osmanthus wine

Dipping sauce
1 tsp cooking oil
80 g red tarocurd
80 g water
60 g sugar
5 g chicken bouillon powder
5 g Chinese rose wine
30 g peanut butter
*Heat oil, add the ingredients and cook over low heat.

Tips
· Fix the shape of the chicken with metal skewers.
· You may use the same marinade on similar recipes, such as chicken wings, squab or duck in red tarocurd sauce. But you have to adjust the marinating time.
· Before deep frying the chicken, prick its skin with a needle, so that moisture can be released and won't cause the bubbles when deep fried.

Spice marinated goose in Chaozhou style

Ingredients
1 black-brown goose

Chaozhou marinade A
10 tbsp goose fat
160 g sliced garlic
160 g sliced shallot
160 g sliced ginger
160 g spring onion (cut into short lengths)
200 g Shaoxing wine

Chaozhou marinade B
8 litres water
12 cloves star anise
4 whole dried tangerine peels
4 tbsp Sichuan peppercorns
2 tbsp five-spice powder
320 g galangal
240 g coriander
80 g Thai basil
2 tbsp red yeast rice
160 g Jinhua ham

Seasoning for marinade
1.6 kg light soy sauce
320 g dark soy sauce
120 g table salt
1.2 kg brown sugar slab
160 g chicken bouillon powder
1.2 kg oyster sauce

Garlic vinegar sauce
8 tbsp Chaoshan rice vinegar
4 tbsp sugar
1/2 tsp table salt
2 tsp grated garlic
1 tsp diced red chilli

Method
1. To make garlic vinegar sauce, put rice vinegar, sugar and table salt in the bowl and stir until dissolve, add garlic and red chilli and mix well.

2. Trim off the solid fat on the inside of the goose. (Save it to be rendered in step 3.) Remove all innards and rinse well. Drain. Cut open the windpipe and the skin of the breast.

3. In a dry wok, heat the goose fat from step 2 over low heat. Save the liquid fat for the marinade. Discard the cracklings.

4. Stir fry the marinade A ingredients until fragrant. That would enhance the flavour of the marinade. Set aside.

5. Boil marinade B ingredients. Turn to low heat and simmer for 1 hour. Turn off the heat. Add seasoning for marinade. Stir until the brown sugar dissolves. Add the marinade A from step 4. This is the Chaozhou marinade.

6. Bring the Chaozhou marinade to a gentle simmer. Put in the dressed goose. Bring to the boil again over high heat. Then turn to low heat and cook the goose for 1 hour until cooked through. Remove from the marinade and let cool.

7. Before serving, chop the goose head, neck and back into pieces. Then de-bone the remaining parts. Slice thinly and arrange on a serving plate. Drizzle with hot Chaozhou marinade. Serve with garlic vinegar sauce.

Tips

· When you cook the duck in the marinade, line the bottom of the pot with bamboo mat, so that the goose won't stick to the pot and get burnt. (pictures 1 & 2 on p.60)

· Goose fat is rendered from the fatty tissue in a goose. Just rinse the fatty tissue and heat it up over low heat in a dry wok. It adds another layer of meaty flavour to the marinade. (pictures 3-5 on p.60)

· You may break the joint on the goose legs before cooking. The thigh meat will shrink into a ball shape after cooked that way. It will be easier to slice too. (pictures 6 & 7 on p.60)

· You may trim the cut near the goose breast and tail into a V shape. That helps fill up the inside of the goose with marinade. The goose can pick up flavours more easily. (pictures 8-10 on p.60)

Crispy "Pipa" quails

Ingredients
3 quails
3 tbsp unforgettable sauce
 (refer to p.15 for method)
3 bamboo skewers (about 6 inches long)

Basting sauce
3 tbsp Basting sauce B
 (refer to p.14 for method)

Marinade
1.5 kg light soy sauce
300 g water
60 g chicken bouillon powder
20 g table salt
60 g sliced ginger
2 whole dried tangerine peels
6 slices liquorice
6 pieces sand ginger
7 bay leaves
4 cloves star anise
20 g cassia bark
20 g galangal
120 g osmanthus wine (added at last)

Method
1. Bring the marinade to the boil over medium heat. Then turn to low heat and simmer for 10 minutes. Turn off the heat and let cool.

2. Remove the innards of the quails. Rinse well. Pour boiling water over the skin to shrink it until it turns light yellow. Rinse with cold water until the skin isn't sticky any more. Put the quails into the marinade from step 1. Leave them for 30 minutes.

3. Cut the quails open by cutting along the breast bone. Spread the body apart and flatten it by inserting bamboo skewer between the ribs. Rinse the skin and wipe dry. Brush the basting sauce B over the skin evenly. Leave them to air-dry or dry them in an oven at 50°C.

4. Preheat an oven up to 150°C. Place the quails on a wire rack with the skin side down. Brush unforgettable sauce on the inside of the quails. Bake for 5 minutes. Flip them and bake the other side at 100°C for 30 minutes.

5. Prick holes evenly on the skin of the quails with a needle. Heat a wok of oil and deep fry the quails until crispy on the outside. Set aside. Brush unforgettable sauce on the inside again. Chop into pieces. Serve.

Tips
· By pricking holes on the quail skin with a needle or a metal skewer, air won't be trapped under skin when deep fried. The skin will be smoother that way.
· Inserting bamboo skewers on the quail helps keep it shape and reduce shrinkage when fried. Optionally, you can shape it like a Pipa (refer to p.165 for the method of Fried chicken in red tarocurd sauce).

Concubine chicken with dried scallops and ham

Ingredients

1 dressed chicken (about 1.35 kg)

Dried scallop marinade

3.6 litres water
60 g dried plaice (toasted and de-boned)
120 g Jinhua ham
120 g dried shrimps
150 g dried scallops
120 g peeled ginger
60 g galangal
10 slices sand ginger
2 tbsp cumin seeds
1/2 tbsp cloves
10 bay leaves
3 cloves star anise
300 g pork bones

Seasoning for marinade

600 g brown sugar slab
350 g table salt
60 g chicken bouillon powder
45 g Shaoxing wine
30 g osmanthus wine
15 g Chinese rose wine

Dipping sauce

2 tbsp grated ginger
2 tbsp finely chopped spring onion
1/4 tsp table salt
1/4 tsp chicken bouillon powder
3 tbsp smoking hot oil

Method

1. Soak the dried scallops in hot water until soft. Rinse the ham. Bake ham, dried plaice and dried shrimps in an oven until fragrant. Put all "dried scallop marinade" ingredients (expect pork bones) into a muslin bag and tie well.

2. Blanch and drain the pork bones. Put them into a pot of water. Put in the spice bag from step 1. Bring to the boil and turn to low heat. Simmer for 2 hours. Turn off the heat.

3. Add "seasoning for marinade." Stir till the brown sugar dissolves. Leave it to cool. Discard the pork bones.

4. Trim the fat on the insides of the chicken. Remove the lungs, windpipe and innards. Rinse well. Blanch in boiling water. Bring to the boil again. Turn to low heat and cook for 5 minutes. Turn off the heat and cover the lid. Leave the chicken in the hot water for 30 minutes. Drain. Dip the chicken in ice water to chill instantly.

5. Drain and wipe dry. Put the chicken in the dried scallop marinade from step 3. Let it soak for 1 hour. Chop into pieces and serve.

Tips

· Instead of whole dried plaice, you may use 30 g ground dried plaice when you make the marinade. It's just more convenient that way.

· The dried scallop marinade is rather potent. The chicken shouldn't be steeped in it for longer than 1 hour. Or else it might be too salty.

Banquet recipes

169

Roast suckling pig with rosemary

Ingredients

1/2 frozen suckling pig (about 1.6 kg)
12 pieces scorched rice slabs
40 g cucumber
1/2 tsp baking soda

Marinade

3.5 tbsp sugar
2 tbsp table salt
1/4 tsp five-spice powder

Basting sauce

4 tbsp Basting sauce A
(refer to p.13 for method)

Rosemary sauce

80 g water
16 g powdered butter
3 g chicken bouillon powder
3 g sugar
1 g dried rosemary

Tips

· Making a few light incisions on the fleshier parts of the pig helps the marinade to seep through.

· When you brush basting sauce on the skin, make sure you brush it from top to bottom continuously. That's how the pig skin picks up the basting sauce evenly.

· Depending on the size of your oven, you may cut off the pig's head or divide the pig into several pieces and bake each part separately. You may cook the head in spiced marinade and serve it separately.

· When you deep fry the scorched rice slabs, the oil temperature must be higher than normal for the rice to puff up and turn crispy.

Method

1. Thaw the pig and lay it flat with the back side down. Cut along the chin. Remove the throat, windpipe, the brain and the caul fat on both sides.

2. Cut along the backbone of the pig from head to tail without cutting through the skin. Pull the pig from both ends to flatten it.

3. Cut off the four ribs near the head side. Slowly remove the fore shank bones and shoulder blade by cutting along the joints. Make a few light incisions on the fleshy part. Then slice off the meat on the hind legs. Make a few light incisions on the fleshy part. Chop off the trotters, hind feet and tail. Hang the pig up with hooks.

4. Boil a pot of water and blanch the pig for 2 minutes until the skin shrinks. Then rinse in cold water until the skin feels cool and no longer sticky. Drain.

5. Lay the pig flat on a baking try with the skin side down. Mix the marinade well and spread on the insides of the pig (not on the skin). Leave it for 10 minutes. Tilt the baking tray to drain any moisture drawn out of the pig by the marinade. That ensures the skin will not pick up too much moisture and stays crispy.

6. Flip the pig to have the skin side facing up. Rub table salt on the skin evenly and leave it for 2 minutes. Dissolve 1/2 tsp of baking soda in 600 ml of warm water (about 70 °C). Rinse off the salt and grease on the pig skin with the soda solution. Rinse with water again. Wipe dry with towel. Brush the basting sauce A evenly on the skin.

7. Preheat an oven to 250 °C. Put the pig on a wire rack with the back facing up. Put the rack over a baking tray lined with aluminium foil. Bake in the oven with the door propped open halfway for 15 minutes. Turn the heat down to 150 °C and bake for 15 more minutes. Lastly, bake at 100 °C for 1 hour until the pork is cooked through. Leave it to air dry at room temperature.

8. Cook the rosemary sauce until thick. Set aside. Cut the cucumber into thin discs.

9. Heat oil up to 180 °C. Deep fry the scorched rice slabs until puffy and crispy. Drain.

10. Prick holes on the pig's skin with a metal skewer. Deep fry the pig in hot oil until crispy. Remove the skin from the pig and trim off any fat under the skin. De-bone the whole pig. Cut the skin and the meat into 1.5-inch squares.

11. To serve, put fried scorched rice slabs on the bottom of a plate. Arrange sliced cucumber and the meat of the suckling pig on top. Drizzle with 1/2 tbsp of rosemary sauce. Lastly arrange the pig's skin on top. Serve.

Braised abalones in unforgettable sauce

Ingredients

12 live baby abalones (about 45 g each)
200 g unforgettable sauce
(refer to p.15 for method)

Seasoning

170 g water
1 tsp table salt
1/2 tsp sugar
1/2 tsp chicken bouillon powder
2 tsp Chinese rose wine
30 g sliced ginger
2 tbsp chopped shallot
2 tbsp grated garlic

Method

1. Scrub the abalones clean with a brush. Shell them and remove the innards. Blanch the shells in boiling water. Rinse well and set aside.

2. Put the shelled abalones in a strainer. Sprinkle with 2 tbsp of potato starch. Rub evenly on abalones for 2 minutes. Rinse with water.

3. To make the seasoning, boil 170 g of water. Put in ginger and cook over low heat for 5 minutes. Leave it to cool. Add the remaining seasoning. Stir well. Put the abalones in. Leave them for 30 minutes. Drain and wipe dry.

4. Stir unforgettable sauce into the abalones. Mix well and leave them for 30 minutes.

5. Line a baking tray with aluminium foil. Preheat an oven to 240 °C. Bake the abalones for 4 minutes on each side. Arrange the blanched shells on a serving plate. Put one abalone in each shell. Serve

Tips

It's not hard to shell abalones. Just insert a small knife between the flesh and shell of the abalone. Slowly cut through. You may even use a metal spoon to lift the flesh off the shell.

Three-cup crabs

Method

1. Put a crab on a chopping board with the abdomen facing up. Make a cut along the centre without cutting through the carapace. Remove the carapace, gills and innards. Rinse and cut into pieces. Chop each claw into halves. Gently crack them with the flat side of knife. Sprinkle with potato starch evenly.

2. Slice the ginger. Cut spring onion and chillies into short lengths. Separate the leaves and stems of the basil.

3. Heat oil in a wok. Put in ginger, garlic and crab. Cook over medium heat until golden. Drain.

4. Heat a wok and add sesame oil. Stir fry aromatics (except basil) over medium heat until fragrant. Put in the crabs and toss well. Pour in the Three-cup sauce. Mix well and cover the lid. Cook over low heat until the liquid reduces (for about 5 minutes). Add basil and stir in the thickening glaze. Stir until the sauce thickens. Serve.

Ingredients

2 swimmer crabs (about 600 g)
4 tbsp potato starch
2 tbsp sesame oil

Aromatics

30 g spring onion
20 g ginger
16 cloves garlic
2 red chillies
15 g Thai basil

Three-cup sauce (mixed well)

1 cup water
1 tsp sugar
1 tsp dark soy sauce
1 tsp light soy sauce
1 tsp spicy bean sauce
1 tbsp oyster sauce
2 tbsp rice wine
2 tbsp Shaoxing wine

Thickening glaze

2 tbsp water
1 tbsp potato starch

Tips

· The carapace doesn't need to be deep-fried in step 3. It is too thin and may crack if deep fried. Just put it in with the crab pieces in step 4. Cover the lid and cook it over low heat. The steam in the wok will cook it through.

· If you use mud crabs instead of swimmer crabs, you should cook them for 2 to 3 minutes longer. It is because mud crabs have thicker shells and are hardly to cook through.

Seafood recipes

Babylon conches in Chinese wine

Ingredients
600 g marine Babylon conches
1.5 cups water
2 tbsp cooking oil
20 g coriander (cut into short lengths)
20 g spring onion (cut into short lengths)

Aromatics
20 g grated ginger
20 g finely chopped shallot
20 g grated garlic
1 red chilli
1 tbsp Sichuan peppercorns

Seasoning
2 cloves star anise
3 bay leaves
2 perilla leaves
1 tbsp Sa Cha sauce
2 tsp spicy bean sauce
1.5 tbsp peanut butter
1 tbsp premium gold soy sauce
1 tbsp dark soy sauce
2 tsp sugar
1 tsp salt
1 tsp chicken bouillon powder

Chinese wine blend
1/2 tbsp Chinese rose wine
1.5 tbsp Shaoxing wine
1.5 tbsp rice wine

Method
1. Rinse the conches. Cook them in boiling water for 2 minutes. Drain.

2. Heat a wok and add 2 tbsp of oil. Stir fry aromatics until fragrant. Put in the blanched conches. Stir well. Add 1.5 cups of water and seasoning. Cook over low heat for 5 minutes.

3. Add the Chinese wine blend. Bring to the boil and turn off the heat. Cover the lid and leave them in for at least 1 hour. Before serving, add coriander and spring onion. Cook them through and serve.

Tips
Babylon conches have rather firm meat. They have to be soak in seasoned stock for a relatively long time to pick up the flavours. If you serve them right after cooked, they won't taste as good.

Spice-marinated Bai Ye dumplings with shrimp and pork filling

Ingredients

3 sheets Bai Ye
(fresh tofu skin, 9-inch square each)
12 shelled shrimps (keep the tail)
90 g white cabbage (shredded)
1/2 tsp baking soda
2 cups warm water

Filling

100 g ground pork
5 g finely chopped spring onion
3 g grated ginger
3 g grated garlic

Seasoning

1/8 tsp five-spice powder
1.5 tsp cooking oil
1 tsp Kikkoman brand soy sauce
1 tsp Shaoxing wine
2/3 tsp sugar
1/4 tsp table salt
1/2 tsp chicken bouillon powder
1/4 tsp sesame oil

Sealing paste (mixed well)

2.5 tbsp glutinous rice flour
3.5 tbsp hot water

Marinade A

2 tbsp cooking oil
5 g red chilli
5 g sliced ginger
5 g sliced shallot
5 g sliced garlic
1 clove star anise
1.5 tbsp Sichuan peppercorns
1 tbsp Shaoxing wine

Marinade B

450 g water
1.5 tbsp light soy sauce
1/2 tbsp dark soy sauce
1/2 tbsp brown sugar
1/2 tsp table salt
1/2 tsp chicken bouillon powder
5 g Chinese celery
5 g spring onion
5 g coriander
1/2 tsp five-spice powder
1 tbsp Sichuan pepper oil

Thickening glaze

1 tbsp potato starch
2 tbsp water

Method

1. Add 1/2 tsp of baking soda to 2 cups of warm water. Mix well. Put the Bai Ye into the solution for 15 minutes until it turns white. Rinse the Bai Ye in running water for 15 minutes.

2. Mix the filling ingredients. Add seasoning and stir until sticky. Put in the shredded white cabbage. Mix well. Refrigerate for 1 hour.

3. Cut each sheet of Bai Ye into quarters. Make a cut on each quarter. Insert the shrimp tail through the cut. Put 1 tbsp of filling on the shrimp. Wrap the Bai Ye into a small parcel. Seal the seam with sealing paste.

4. Heat wok and add oil. Fry the Bai Ye dumplings over low heat until golden. Drain.

5. In a wok, stir fry marinade A ingredients until fragrant. Add marinade B ingredients and bring to the boil. Put in the Bai Ye dumplings. Bring to the boil again. Turn to low heat. Cover the lid and cook for 5 minutes.

6. Arrange the Bai Ye dumplings on a serving plate. Stir the thickening glaze into the marinade. Drizzle over the dumplings. Serve.

Tips

Soaking the Bai Ye in baking soda solution not only makes it looks whiter, but also softens it texture.

Lobster in Kung Pao cheese sauce with Udon noodles

Ingredients

1 lobster (about 900 g)
2 bundles of Udon noodles
4 strips smoked bacon (diced)

Kung Pao cheese sauce

8 slices cheddar cheese
60 g butter
3 tbsp flour
3 tsp sugar
1 tsp chicken bouillon powder
1 tbsp spicy bean sauce
1/2 onion (diced)
100 g green and red bell pepper (diced)
2 cups water

Stock mix

1 cup stock
1/3 tsp salt
1/2 cup Kung Pao cheese sauce

Method

1. To make the Kung Pao cheese sauce, melt butter in a pan over low heat. Stir fry diced onion until fragrant. Sprinkle with flour and stir into thick paste known as a roux. Slowly stir in water to thin it out. Add all other ingredients (except the bell peppers). Cook over low heat until the cheese melts. Stir in the bell peppers.

2. To prepare the lobster, cut it in half along the length on the belly side. Rinse and cut into pieces. Drain. Sprinkle with about 1 tbsp of potato starch. Pat the spread evenly. Heat a wok and add oil. Fry the lobsters on both sides until done and lightly browned.

3. Blanch Udon noodles in water and stir to scatter them. Drain. Put the stock mix in a pot. Bring to the boil and put in the blanched Udon. Mix well. Cook over medium heat until the Udon is tender, but not overcooked (for about 2 minutes). Transfer the noodles and the sauce into a baking tray.

4. Arrange the lobster over the Udon. Pour the remaining Kung Pao cheese sauce over the lobsters. Sprinkle with diced bacon on top.

5. Preheat an oven to 240 °C. Bake the lobster for 10 minutes until golden on the surface. Serve.

Tips

When you dress the lobster, pierce the tail along the length with a wooden chopstick. This helps release the metabolic waste in the lobster, so that it won't taste like ammonia after cooked.

Wine-marinated mantis shrimps

Ingredients
2 giant mantis shrimps (about 250 g each)
4 tbsp Shaoxing wine

Marinade
600 g water
15 g sliced ginger
15 g spring onion (cut into short lengths)
2 tsp table salt
1 tsp chicken bouillon powder
1/2 tsp five-spice powder
5 tbsp Shaoxing wine (added at last)

Osmanthus wine blend
1 tbsp Shaoxing wine
1 tbsp osmanthus wine
1/2 tbsp candied osmanthus

Method

1. Boil the marinade. Turn to low heat and simmer for 10 minutes. Turn off the heat and leave it to cool. Stir in Shaoxing wine. Refrigerate until chilled.

2. Rinse the mantis shrimps and drain. Soak the mantis shrimps in 4 tbsp of Shaoxing wine for 15 minutes.

3. Boil water in a steamer. Steam the mantis shrimps for 4 minutes. Dunk into the chilled marinade from step 1.

4. Cut along the back of the mantis shrimps with scissors. Put them back in the marinade. Refrigerate the shrimps in the marinade for 1 hour.

5. Before serving, shell the shrimps and chop them up into pieces. Save on a serving plate and drizzle with the osmanthus wine blend.

Tips

In step 2, soaking the mantis shrimps in Shaoxing wine intoxicates them so that they release their metabolic waste readily. It also adds a wine fragrance to the shrimps and crisps up the flesh.

Crispy honey-glazed eel

Ingredients

300 g white eel (de-boned)
1/2 tsp baking soda
2 tbsp honey

Basting sauce

20 ml Basting sauce A
(refer to p.13 for method)

Marinade

5 tsp sugar
1 tsp table salt
1 tsp chicken bouillon powder
2 tsp Hoi Sin sauce
2 tsp oyster sauce
1 tsp satay sauce
1 tsp dark soy sauce
1 tsp Chinese rose wine
1 tsp Shaoxing wine
2 tsp finely chopped shallot
2 tsp grated garlic
1 tbsp shredded ginger

Method

1. Soak the eel in hot water briefly. Scrub off the slime on the skin. Cut and discard the head and tail. Cut along the back of the eel. Remove the backbone and innards. Open it up into one piece of fillet. Rinse and drain.

2. Lay the eel flat with the skin side up. Prick holes on the skin evenly with a skin pricker. Rub baking soda on the skin and leave it for 5 minutes. Rinse well.

3. Mix the marinade. Put the eel in and leave it for 40 minutes. Rinse the marinade off the skin. Insert a bamboo skewer into the fleshier part of the eel. Hold the hook to hang the eel. Pour boiling water on the eel until the skin shrinks. Wipe dry. Brush basting sauce A on it. Leave it to air dry.

3. Heat oil in a wok up to 200 °C. Turn to low heat. Put in the eel. Shallow-fry the eel with the skin side down until medium-well done. Flip it to fry the other side until golden. Brush honey on the skin. Slice diagonally and save on a serving plate. Serve.

Tips

You may ask the fishmonger to do step 1 for you. That'll save you some time and effort.

Garlic-scented crispy roast pork

Ingredients
1.2 kg boneless pork belly (skin-on)
80 g sliced ginger
80 g spring onion

Method

1. Rinse the pork belly. Blanch in a pot of water with sliced ginger and spring onion (put the bamboo mat on the bottom of the pot). Bring to the boil again and cook over low heat for 20 minutes. Rinse in cold water until cool to the touch and the skin no longer sticky. Drain.

2. Lay the pork belly flat with the skin side down. Make light one incision at one-inch intervals. Rub the marinade on the meat and into the creases. Leave it for 1 hour 30 minutes.

3. Flip the pork upside down (skin side up). Scrape off any hair and impurities on the skin. Prick holes on the skin evenly with a skin pricker (without going too deep into the subcutaneous fat). Brush the basting sauce over the skin. Leave it for 30 minutes. Scrape off any salt crystals remaining on the skin.

4. Preheat an oven to 200 °C. Put the pork on a wire rack with the skin side up. Put the rack on a baking tray lined with aluminium foil. Bake the pork with the door ajar for 10 minutes until the pork skin looks dry. Take the pork out to prick more holes on the skin with a skin pricker (again without going too deep into the fat layer).

5. Keep the oven at 200 °C. Bake the pork belly with the skin side up for 30 more minutes. Wrap the pork in aluminium foil on the vertical sides, exposing the skin. Turn up the oven to 250 °C. Bake for 30 minutes until the skin puffs up and turn crispy. Chop into pieces and serve with garlic dip.

Marinade
3 tbsp table salt
2 tbsp sugar
1 tbsp chicken bouillon powder
1/2 tsp five-spice powder
2 tbsp deep-fried grated garlic
1 tbsp raw grated garlic

Basting sauce
1 tbsp white vinegar
1/2 tbsp Chinese rose wine
1 tbsp table salt (added at last)

Garlic dip (mixed well)
1 tbsp yellow mustard
1 tbsp unforgettable sauce
(refer to p.15 for method)
1 tsp deep-fried grated garlic

Tips
· After baking for a while, the pork skin may show dark burnt spots. Just scrape them off with a paring knife and keep baking until the skin turns granular with small bumps.

· The pork skin may show big bumps at certain spots. That is because of the uneven heat that the skin is exposed to. You may move the wire rack upward to get closer to the top heat filament in your oven. Keep baking until the skin turns granular with small bumps.

Festive favourites

New Year shredded salad with abalone and jellyfish

Ingredients

200 g jellyfish
100 g roast goose (de-boned)
100 g concubine chicken (de-boned)
1 can instant baby abalones
100 g celery
100 g carrot
120 g cantaloupe
120 g Hami melon
60 g corn flakes

Marinade

1/2 tsp table salt
1 tsp sugar
1/2 tsp chicken bouillon powder
2 tsp sesame oil

Dressing

80 g Japanese sesame dressing
80 g Thai sweet chilli sauce
2 tsp sesame oil

Method

1. Mix the dressing and keep in the fridge.

2. Brush off the salt on the jellyfish. Roll it into a cylinder. Shred it (about the thickness of a chopstick).

3. Heat water (without boiling). Put in the shredded jellyfish and cook until it shrinks. Drain. Rinse in cold water until it swells again and no salty (for about 2 hours). Drain. Refrigerate.

4. Shred the chicken and goose. Slice the abalones. Peel and shred the melons.

5. Rinse celery and carrot. Shred them and blanch in boiling water briefly. Drain and soak in ice water until cool. Drain.

6. Mix the jellyfish with the marinade. Leave it for 10 minutes. Drain.

7. On a large serving plate, put the jellyfish at the centre. Arrange the remaining ingredients in group around the jellyfish. Top with sliced abalones. Drizzle with the dressing. Before eating, invite guests to toss the salad together with chopsticks. This has a lucky connotation of prosperity and liveliness.

Tips

· Do not blanch the jellyfish in boiling water. The water should be 80°C. Otherwise, the jellyfish will be overcooked and fail to shrink and swell. It won't taste as springy that way.

· For the recipe of Concubine chicken, please refer to p.169.

Eight-treasure stuffed dace

Method

1. Steam the dried pork, pork sausage and liver sausage for 6 minutes. Dice finely.

2. Soak shiitake mushrooms, dried shrimps and dried tangerine peel in water until soft. Dice them finely. Dice coriander and spring onion finely.

3. In a mixing bowl, put in the minced dace. Stir until sticky. Add seasoning and stir in one direction to mix well. Stir vigorously until sticky. Lift it up and slab it hard into the bowl for 10-plus times.

4. Add all remaining Eight-treasure minced dace filling ingredients. Stir well and refrigerate for 30 minutes.

5. Dust the inside of the de-boned dace with potato starch. Stuff it with the filling from step 4. Press the filling to compact it and smooth the surface to resemble the belly of a fish. Sprinkle potato starch on the fish skin.

6. Heat oil in a wok and fry the stuffed fish over medium heat until half done on all sides. Set aside. Steam for 15 minutes until cooked through. Cut into pieces. Arrange on a plate.

7. Stir fry aromatics in a little oil until fragrant. Add stock and sauce. Bring to the boil and pour over the stuffed dace. Serve.

Ingredients

1 dace (about 400 g, de-boned, filleted with the skin on)

Aromatics

diced spring onion
diced shiitake mushrooms
grated garlic

Eight-treasure minced dace filling

400 g minced dace
30 g dried shrimps
30 g dried pork
20 g dried Chinese pork sausage
20 g dried Chinese liver sausage
30 g shiitake mushrooms
20 g coriander, spring onion and dried tangerine peel

Seasoning for filling

1 tsp salt
1/2 tsp chicken bouillon powder
1 tsp sugar
1 tsp sesame oil
2 tbsp potato starch
4 tbsp water
ground white pepper

Sauce

1/3 tsp salt
1 tbsp oyster sauce
1 tsp dark soy sauce
2 tsp potato starch
1/2 cup stock

Tips

· You can ask the fishmongers to de-bone the dace while keeping the skin intact. Just tell them you're going to make stuffed dace.

· The dace flesh you get from one fish isn't enough to stuff it plump and full. Thus, you have to buy extra minced dace to add to it. You may also get seasoned minced dace from the market. Just make sure you cut the seasoning to half if you do.

· After mixing the filling, I refrigerate it for 30 minutes. That helps firm it up a little so that it's easier to stuff the fish with.

· Dusting the inside of the dace with potato starch helps the filling to bond better. It is less likely to fall apart when you slice it. Besides, make sure you stuff more filling towards the head and tail and press firmly. The skin will bond better with the filling that way.

· After stuffing the dace with filling, it might deform and look odd. Shape it nicely with your hands before frying.

· For best results, stuff the fish and fry it right away. Yet, you may make the filling one day ahead and keep it in the fridge.

Fish toasts with banana

Method

1. Cut the fish and sandwich bread into rectangles of 1 inch x 2 inches. Add marinade to the fish. Mix well.

2. Whisk the eggs and potato starch.

3. Cut banana and ham into small pieces. Add a little egg wash and mix well.

4. Spread some butter on a piece of bread. Put a piece of banana on it. Dunk a piece of fish into the egg wash. Put it over the banana. Arrange ham and coriander on top. Arrange neatly on a lined baking tray.

5. Preheat an oven to 200 °C for 10 minutes. Bake the fish toasts for 10 minutes until the bread turns crispy and golden. Serve.

Ingredients

3 pieces fish fillet
2 slices sandwich bread
1 banana
1 slice cooked ham
30 g butter
2 sprigs coriander

Marinade

1/2 tsp salt
1/3 tsp chicken bouillon powder
sesame oil
ground white pepper
2 tsp potato starch

Egg wash

2 eggs (whisked)
8 tbsp potato starch

Tips

· If you don't have an oven, you can deep-fry the toasts instead. Just pay attention to the oil temperature in that case. The oil should be 200℃ when you put in the fish toasts and it takes about 4 minutes to turn the bread golden. If the oil is any cooler, the bread will be soaked with oil and becomes disgustingly greasy.

· Optionally, serve with tartar sauce, creamy salad dressing or sweet and sour sauce on the side.

Festive favourites

Puff pastry mooncake with barbecue pork filling

Water dough
150 g cake flour
40 g shortening
20 g sugar
1/2 egg
20 g custard powder
70 ml water

Oil dough
80 g cake flour
70 g shortening

Method

Water dough

1. Mix sugar, egg and shortening until well blended. Stir in flour and custard powder a little at a time. Then add a little water. Keep repeating with flour and water until smooth.

2 Roll the dough out into a rectangular. Put it in a rectangular tray and wrap in cling film. Leave it to rest for 1 hour. Refrigerate for 1 hour.

Oil dough

Mix shortening with flour until smooth. Press into a rectangular tray. Wrap in cling film and refrigerate for 2 hours.

Puff pastry dough assembly

1. Roll the water dough out so that it is 3 times wider than the oil dough.

2. Lay the water dough flat on the counter. Put the oil dough over it at the centre. Fold the water dough from both sides to wrap the oil dough. Roll it out with rolling pin. Fold in thirds again. Roll into a rectangle. Refrigerate for 1 hour.

3. Fold the dough in thirds. Roll it out into a rectangle. Wrap in cling film and refrigerate for 2 hours.

Tips

· When you make the puff pastry, you must refrigerate the dough for an hour after pressing it into rectangle. Otherwise, the shortening in the dough might melt and the layering of the puff pastry becomes impossible.

· When you wrap the filling in the dough, use care not to bring the dough. Otherwise, the cake will burst when baked.

Barbecue pork filling

200 g barbecue pork
50 g onion
2 salted egg yolks (each cut into quarters)
10 g white sesames
2 tbsp multi-floral honey
1 egg (whisked)

Seasoning

2 tbsp oyster sauce
1 tbsp light soy sauce
1 tbsp dark soy sauce
4 tbsp sugar

Thickening glaze (mixed well)

50 ml water
60 g cornstarch

Method

Filling

1. Finely dice barbecue pork and onion.

2. Heat a wok and add 3 tbsp of oil. Stir fry onion and barbecue pork until fragrant. Add 200 ml of water. Bring to the boil. Add seasoning and cook until sugar dissolves. Thicken with cornstarch slurry. Cook until ingredients are done. Leave it to cool and refrigerate for 1 hour.

Mooncake Assembly

1. Preheat an oven to 200 °C for 10 minutes.

2. Roll the puff pastry dough out. Roll it out to 1/2 cm thick. Cut out round disc with a 3-inch cookie cutter.

3. On each disc of dough, put 1 tbsp of filling and 1/4 salted egg yolk. Fold the dough up and seal the seam. Shape it into a round thick patty. Arrange on a lined baking tray. Brush egg wash over the mooncakes.

4. Bake in the oven for 15 minutes. Brush on one more coat of egg wash and sprinkle with white sesames. Turn the oven down to 180 °C. Bake for 5 minutes. Turn off the oven. Leave the mooncakes inside for 2 minutes. Remove and brush on the honey evenly. Let them cool. Serve.

Grilled spareribs in Peking sauce

Ingredients

1 rack spareribs (about 900 g)
1 sprig Peking scallion (sliced)
1 onion (shredded)

Seasoning

8 cups water
4 tbsp ketchup
2 tbsp OK sauce (steak sauce)
1 tbsp Worcestershire sauce
2 tbsp spiced vinegar
3 tbsp Maggi's seasoning
4 tbsp oyster sauce
2 tsp salt
2 tsp chicken bouillon powder
200 g rock sugar
20 g red yeast rice
50 g ginger and spring onion
40 g spices (Tsaoko fruit, bay leaves, star anise)

Method

1. Brush light soy sauce on the spareribs. Heat 1 tbsp of oil in a wok and sear the spareribs over medium heat until lightly browned.

2. Put the spareribs into a pot. Add seasoning (and water or stock to top up if not enough). Cook over medium low heat for 1.5 to 2 hours until tender. Wait till it cools slightly and de-bone them.

3. In a baking tray, put in shredded onion. Put the de-boned spareribs over the bed of onion. Add 1 cup of the braising stock from step 2. Bake in an oven at 220 °C for 15 minutes. Save on a serving plate.

4. Heat up the braising stock from step 2. Thicken with potato starch slurry. Drizzle over the spareribs. Stir fry Peking scallion in a little oil. Put Peking scallion on top. Serve.

Tips

· Before baking the spareribs in oven, it's advisable to pour some braising stock over it. Otherwise, the spareribs could be too dry and too burnt.

· To heighten the flavour of the braising stock, you may reduce it further over medium heat after the spareribs are done.

· Optionally, you may serve this dish with steamed and fried buns which taste divine after dipped into the rich flavourful sauce.

Glutinous rice dumplings with marinated pork filling

Ingredients

600 g glutinous rice
150 g pork belly
80 g black-eyed beans
5 salted egg yolks
10 chestnuts
10 ready-to-serve abalones
(24 pcs in 600 g, i.e. 25 g each)
10 dried shiitake mushrooms
1 tbsp rice wine
30 large bamboo leaves
10 segments cotton strings
(each about 1 m long)

Marinade A

4 tbsp cooking oil (added first)
15 g sliced ginger
80 g shallots
2 cloves star anise
1 tbsp Shaoxing wine (added at last)

Marinade B

600 g water
1.5 tbsp light soy sauce
1 tbsp dark soy sauce
1 tbsp brown sugar
1/2 tbsp table salt
1 tbsp oyster sauce
1/2 tbsp chicken bouillon powder
1/2 tsp five-spice powder
4 bay leaves

Tips

· Mixing the marinade with the glutinous rice adds a caramel colour to the dumplings and gives extra fragrance of spices.

· I steam the glutinous rice first to partly cook it before wrapping in bamboo leaves. That helps shorten the final steaming time.

· Baking the salted egg yolks in oven enhances the flavours.

· When you wrap the glutinous rice in bamboo leaves make sure you get the folding and overlapping parts in the right places. Otherwise, the dumplings might leak.

· You may hammer a nail on the wall and tie one end of the cotton string on it before wrapping the dumplings. That will make your life a lot easier.

Method

1. Soak shiitake mushrooms in water until soft. Steam until done (refer to p.17 for method).

2. Soak the dried bamboo leaves in water until soft. Scrub well and rinse. Blanch in boiling water until they turn bright green. Drain. Soak in cold water for later use.

3. Soak the glutinous rice in water for 5 hours. Drain. Steam for 30 minutes. Wrap it up in cheese cloth.

4. Soak the black-eyed beans in water for 5 hours. Drain. Shell and peel the chestnuts. Rinse well.

5. Add 1 tbsp of rice wine to the salted egg yolks. Leave them for 5 minutes. Bake in an oven at 150°C for 5 minutes. Leave them to cool. Cut them into halves.

6. Cut the pork belly in 10 pieces. Each about 15 g.

7. Heat a wok and stir fry marinade A ingredients until fragrant. Add marinade B. Bring to the boil. Put a bamboo mat on the bottom of the pot. Put in shiitake mushrooms, black-eyed beans, chestnuts and pork belly. Bring to the boil and turn to low heat. Cover the lid and simmer for 45 minutes. Remove all ingredients from the marinade.

8. Put the steamed glutinous rice in a big mixing bowl. Add 200 g of the marinade from step 7. Put in the cooked black-eyed beans. Mix well. Divide into 10 equal portions.

9. To wrap the dumpling in bamboo leaves, lay 2 leaves flat with the smooth side up. Overlap half of the leaves along the length. Fold in half to make ends meet. Roll one end into the fold while holding the fold tightly to create a pouch. Put in half portion of glutinous rice. Arrange shiitake mushrooms, salted egg yolk, chestnut, pork belly and abalone over the rice. Top with the remaining half portion of rice.

10. Take one bamboo leaf to cover the open end of the pouch. Fold about 1 inch inward to seal well. Turn the pointy tip of the pouch upward and the lastly added leaf downward. Wrap the dumpling tightly with cotton string. Steam the dumplings for 40 minutes. Serve.

Glutinous rice balls with pork and sesame filling

Method

Dough

1. In a mixing bowl, put in 4 tbsp of glutinous rice flour. Add 2 tbsp of water. Knead into dough. Boil the dough in a pot of water until it floats.

2. Put the cooked dough into a big pot. Add 310 g glutinous rice flour, 250 g warm water and 1.5 tbsp cooking oil. Knead into dough. Set aside.

Assembly

1. Boil the stock for 15 minutes over low heat.

2. Soak shiitake mushrooms in water until soft. Rinse and steam (refer to p.17 for method).

3. Soak the tribute vegetable in water for 30 minutes to make it less salty.

4. Cook the pork cheek in Chaozhou marinade (see recipe of Marinated Goose in Chaozhou Style on p.166 for method) over low heat for 30 minutes. Set aside.

5. Dice finely the shiitake mushrooms, tribute vegetable and pork cheek.

6. Toast the white sesames in a dry wok over low heat until golden.

7. Heat a wok and add oil. Stir fry shallot over low heat until fragrant. Add pork cheek, tribute vegetable and shiitake mushrooms. Stir well. Sizzle with wine. Add dark soy sauce and oyster sauce. Thicken with potato starch slurry. Turn off the heat. Sprinkle with toasted sesames. This is the filling.

8. Get 30 g of dough. Roll into a sphere. Flatten it. Wrap 1 tbsp of filling in the dough. Seal the seam and roll into a sphere.

9. Boil a pot of stock. Put in the glutinous rice balls. Cook over medium heat for 5 minutes until they float. Drain and serve.

Ingredients for dough

350 g glutinous rice flour
250 g warm water (about 40 °C)
1.5 tbsp cooking oil
2 tbsp water

Stock

2 cups chicken stock
2 cups water
1 tsp table salt
15 g sliced ginger

Filling

1 tbsp cooking oil
40 g chopped shallot
120 g pork cheek
50 g tribute vegetable (dried celtuce stems)
50 g shiitake mushrooms
1 tbsp Shaoxing wine
1 tsp dark soy sauce
1 tsp oyster sauce
2 tbsp white sesames

Thickening glaze (mixed well)

2 tbsp water
2 tsp potato starch

Tips

· To make the dough more springy and finer in texture, I cook part of the glutinous rice dough first and then mix it with uncooked flour.

· When you wrap the filling with the dough, you should flatten the dough so that it is thicker at the centre and thinner on the rim. The filling is less likely to leak and the dumplings are less likely to break this way.

· If you want to save time and effort, you may buy readymade marinated pork belly from stores, instead of making your own pork cheek from scratch.

Jellied pork cold cut

Ingredients

2 salted pork trotters
5 fresh chicken feet
300 g pork skin
30 g sliced ginger
3 tbsp Chinese rose wine
1 tbsp sesame oil
30 g gelatine powder (added at last)

Spices

2 slices liquorice
1 clove star anise
4 bay leaves
2 Tsaoko fruits
1 g cassia bark
5 slices sand ginger
8 cloves
1 tbsp cumin seeds
30 g sliced ginger
30 g spring onion

Marinade base

10 cups water
20 g sugar
40 g table salt
10 g chicken bouillon powder
20 g Shaoxing wine
1/2 tsp ground white pepper

Dipping sauce

3 tbsp Zhenjiang black vinegar
10 g shredded ginger

Method

1. To make the marinade, put the spices into a muslin bag and tie well. Boil the marinade base and put in the spices. Cook over low heat for 30 minutes.

2. Burn off the hair on the pork trotters and pork skin with a kitchen torch. Clip off the nails on the chicken feet. Rinse pork trotters, pork skin and chicken feet well. Blanch them in boiling water. Add 30 g sliced ginger and 3 tbsp of Chinese rose wine to the water. Cook for 5 minutes. Drain and rinse with water.

3. Put the pork trotter, pork skin and chicken feet into the spiced marinade from step 1. Bring to the boil. Turn to low heat and simmer for 3 hours. Take the pork trotters out. Skin and de-bone them. Dice the meat coarsely. Keep the spiced marinade for later use.

4. Add gelatine powder to 150 ml of water. Stir until it dissolves (or warm it up over a pot of simmering water to speed up the process). Add the gelatine solution to the spiced marinade. Mix well. Put in pork skin and diced pork and soak for 5 minutes. Set aside.

5. Grease a rectangular tray with a little sesame oil. Put a layer of pork skin on the bottom. Top with a layer of diced pork trotter meat. Add the gelatine-marinade mixture to cover the pork. Top with another layer of pork skin.

6. Use another rectangular tray as a weight on top to keep the surface flat. Refrigerate for 4 hours until set. Cut into cubes and serve the dipping sauce on the side.

Tips

To ensure all ingredients bond well, I suggest putting extra weight when you put another rectangular tray over the jellied cold cut in the last step. Otherwise, the meat may separate from the jelly and pork skin.

Duck tongues in spiced marinade

Ingredients
400 g frozen duck tongues

Aromatics
20 g diced garlic
20 g diced shallot
20 g diced ginger
2 tbsp cooking oil
1 tbsp Shaoxing wine

Spiced marinade
760 g water
130 g light soy sauce
150 g oyster sauce
30 g dark soy sauce
5 g table salt
100 g brown sugar slab
30 g chicken bouillon powder
2 cloves star anise
1 whole dried tangerine peel
5 slices liquorice
5 g Sichuan peppercorns
10 g cumin seeds
10 g cloves

Method

1. Heat wok and add oil. Stir fry aromatics until fragrant. Add spiced marinade ingredients. Bring to the boil over medium heat. Turn to low heat and simmer for 10 minutes. Turn off the heat and cover the lid. Leave the ingredients to steep in the hot marinade for flavours to infuse. When the marinade is cooled completely. Strain and set aside.

2. Thaw the duck tongues and blanch in boiling water. Bring to the boil again. Turn off the heat and let them soak in the hot water for 10 minutes. Rinse with cold water until cooled and no longer sticky. Drain.

3. Put the duck tongues from step 2 into the marinade from step 1. Refrigerate for 2 hours. Serve.

Tips
Do not cut off the bones on the duck tongues, or else they will shrink and don't look good.

Stuffed chicken wings wrapped in pork belly

Ingredients

6 frozen whole chicken wings
300 g skinless whole pork belly
60 g carrot
60 g celery
60 g Peking scallion
30 g coriander
300 g multi-floral honey

Marinade

12 tbsp sugar
2 tbsp table salt
1 tbsp chicken bouillon powder
4 tbsp Hoi Sin sauce
4 tbsp oyster sauce
1 tbsp sesame paste
1/2 tbsp light soy sauce
1/2 tbsp dark soy sauce
3 tbsp Shaoxing wine
2 tbsp grated garlic
1 tbsp finely chopped shallot
2 eggs

Tips

· De-boned wings and thinly sliced pork belly pick up seasoning quickly. Thus, do not leave them in the marinade for too long. Otherwise, they may be too salty.

· When you wrap the wings in pork belly, if one slice of pork isn't long enough to wrap around the full length of the wing, you can continue with one more slice of pork.

Method

1. Lay the pork belly flat on a tray. Leave it in the freezer until stiff. Slice into 6 thin slices. Thaw them at room temperature.

2. Make a cut on the chicken wings joints and then fold the flesh backward to remove the bones in the drummette and mid-joint. Leave the bones in the wing tips. Rinse and drain. Add marinade and mix well. Leave them for 3 minutes.

3. Remove the leaves of coriander and use the stems only. Peel the carrot. Remove the tough veins on the celery with a fruit peeler. Rinse the Peking scallion. Cut carrot, coriander, celery and Peking scallion into thick strips about 5 inches long.

4. Put the pork belly slices into the marinade. Mix well.

5. Stuff each chicken wing with 5 strips of coriander stems, 10 strips of celery, 10 strips of carrot and some shredded Peking scallion. Wrap the stuff wing with a slice of pork belly starting from the wing tip at an angle. Secure the ends of the pork belly by inserting a toothpick.

6. Preheat an oven up to 240 °C. Arrange the stuffed wings flat on a wire rack. Put the rack over a baking tray lined with aluminium foil. Bake for 5 minutes. Flip the stuffed wings to grill the other side for 5 more minutes. Turn the heat down to 125 °C and bake for 10 minutes.

7. Brush honey on the stuffed wings. Bake in the oven at 240 °C for 3 minutes. Flip them and bake for 3 more minutes. Turn the heat down to 125 °C. Bake for 10 minutes. Brush honey on the stuffed wings again. Serve.

Smoked fish in sweet marinade

Method

1. To make the sweet marinade, heat a wok and put in cooking oil. Stir fry the aromatics and spices until fragrant. Add the sweet marinade base. Bring to the boil and turn to low heat. Simmer for 30 minutes. Leave it to cool. Strain and set aside.

2. Rinse the fish and cut into 2-cm thick pieces.

3. Cook the marinade A over low heat for 10 minutes. Leave it to cool. Put in marinade B and let the ingredients soak for 40 minutes. Put in the fish and leave it for 25 minutes. Drain with a strainer. Wipe the fish dry completely with paper towel.

4. Heat oil up to 200 °C. Deep fry the fish for about 1 minute. Turn to low heat and deep fry until golden. Drain. Heat up the oil further over high heat. Deep fry the fish once more until crispy on the surface. Drain.

5. Put the deep-fried fish into the sweet marinade from step 1. Cook over low heat for 5 minutes. Drain and leave the fish to cool.

6. Arrange the fish pieces on a serving plate. Pour in the sweet marinade. Let the fish soak in it for 2 hours until the fish picks up the flavours. Serve.

Tips

· Do not slice the fish too thinly. Otherwise, the fish may break down into bits or dry out in the deep frying process.

· Deep-frying the fish twice can remove the fishy smell. It also dries up the surface of the fish, so that it picks up the sweet marinade more easily.

Ingredients

600 g grass carp belly

Marinade A

450 g water
40 g sliced ginger
1/4 tsp ground white pepper
15 g table salt
10 g sugar
5 g chicken bouillon powder
10 g fish sauce

Marinade B

10 g garlic cloves
30 g shallot
20 g rice wine

Aromatics and spices

2 tbsp cooking oil
20 g sliced ginger
20 g spring onion
20 g shallot cloves
3 cloves star anisc
1 tsp Sichuan peppercorns
1 tbsp rice wine (added at last)

Sweet marinade base

600 g water
50 g golden syrup
55 g teriyaki sauce (1 pack)
60 g Hoi Sin sauce
20 g oyster sauce
1 tsp table salt
1 tsp chicken bouillon powder
1 tsp dark soy sauce
1 tsp Chaoshan rice vinegar
5 bay leaves
1 g cassia bark
6 pieces sand ginger
1/4 tsp ground white pepper
1 tsp chilli oil

Smoked vegetarian beancurd skin roll

Ingredients
2 large sheets beancurd skin
2 tsp sesame oil
1 sheet aluminium foil

Filling
2 tbsp cooking oil
120 g shredded shiitake mushrooms (refer to
p.17 for method)
100 g celery
100 g carrot
1 piece wood ear fungus

Seasoning
120 g boiled water
1 tsp sugar
1/2 tsp table salt
2 tsp oyster sauce
1/2 tsp five-spice powder
2 tsp potato starch

Smoking chips
20 g tea leaves
20 g flour
30 g sugar

Method

1. Soak the wood ear fungus in water until soft. Finely shred. Rinse carrot and celery. Peel and finely shred them.

2. Mix the seasoning well. Set aside 40 g of the mixture.

3. Heat oil in wok and stir fry the filling until fragrant. Add 90 g of the seasoning mixture from step 2. Stir until it dries up. Set aside to let cool.

4. Trim the dry hard rims around the beancurd skin. Brush the 40 g of seasoning mixture from step 2 on both sides of the beancurd skin. Fold it into a semi-circle. Put the stir-fried filling over the beancurd skin. Fold and roll it up into a rectangular parcel about 2 inches wide.

5. Brush sesame oil over the beancurd skin roll. Steam in a steamer for 5 minutes. Leave it to cool.

6. Heat a wok and add 2 tbsp of oil. Fry the beancurd skin roll over low heat until crispy and golden. Set aside.

7. Line the wok with aluminium foil. Spread the smoking chips evenly over it. Put a wire rack over the smoking chips and put the fried beancurd skin roll on top. Turn on medium heat and cook until smoke appears. Cover the lid. Turn to low heat and keep cooking for 2 more minutes. Turn off the heat and leave the lid on. Keep the beancurd skin roll in for 2 more minutes. Brush sesame oil on the beancurd skin roll. Cut into pieces. Serve.

Tips
· When you make the beancurd skin roll, make sure you spread the filling evenly and roll the beancurd skin tightly. Otherwise, the filling may leak in the frying and smoking process.
· Before smoking anything in a wok, make sure you line it with aluminium foil. Otherwise, the smoking chips may burn and char, making it a hard job to clean the wok.
· Adding flour to the smoking chips gives more complexity to the smoked food. Wait till the sugar starts to melt before you put in the beancurd skin roll.

Grilled homemade pork and foie gras sausage

Method

1. Soak the sausage casing in water until soft. Cut into three equal lengths (about 50 cm each). Cut the strings into three segments, each 20 cm long, to be used to tie the sausage ends.

2. Mix the marinade.

3. Finely dice the shiitake mushrooms, water chestnuts, pork and foie gras. Add coriander, potato starch and egg. Mix well. Add 2 tbsp of marinade and stir well. Leave it for 5 minutes. This is the sausage filling.

4. Tie a knot on one end of the casing segments. Insert a funnel from the other end. Fill the casing with filling until full. Tie the open end with a cotton string. Brush marinade on the casing and leave it for 5 minutes.

5. Preheat an oven up to 150°C. Put the sausages flat on a baking tray lined with aluminium foil. Prick holes around the ends of the sausages with a fine needle. That helps release the trapped air in the sausage when cooked.

6. Bake the sausages for 15 minutes. Turn the heat down to 100°C. Flip the sausages and bake for 15 more minutes. Serve with dipping sauce on the side.

Ingredients
300 g half-fatty pork shoulder butt
6 cooked shiitake mushrooms
 (refer to p.17 for method)
80 g water chestnuts (peeled)
80 g foie gras
1 tbsp diced coriander
2 tsp potato starch
1 egg

Casing and utensils
150 cm artificial sausage casing
60 cm cotton string
1 funnel

Marinade
5 tsp sugar
1 tsp table salt
1/2 tsp chicken bouillon powder
2 tsp Hoi Sin sauce
2 tsp oyster sauce
1 tsp sesame paste
1 tsp light soy sauce
1/2 tsp dark soy sauce
1 tsp grated garlic
3 tsp finely chopped shallot
1 tsp Chinese rose wine

Dipping sauce
2 tbsp sweet plum sauce
2 tbsp candied osmanthus

Tips

· When you fill the casing with filling, air may be trapped and stop the filling from reach the end. If that happens, prick 2 to 3 holes on knot of the casing. The air can be released and the filling can reach the end that way.

· You can get artificial sausage casing from the shop producing homemade preserved sausage in Sai Wan.

Snack and appetizers

Lettuce wrap with beef shin and flying fish roe

Method

1. Boil the soaking stock. Turn off the heat and leave it to cool.

2. Slice the beef shin thinly, about 1.5 inch X 3 inches each. Cut each sheet of Nori seaweed into 6-inch square. Shred the young ginger pickle coarsely.

3. Trim each Romaine lettuce leaf into a leaf shape. Slice the cucumber into round thick slices. Scoop out the seeds and pith at the centre to form a hollow tube.

4. Cut one sheet of beancurd skin into 4 pieces, each 6 inches X 6 inches. Soak the beancurd skin into the soaking stock from step 1 until soft and turn bright yellow. Set aside.

5. Put the other sheet of beancurd skin into the soaking stock in whole until soft. Lay it flat and fold into a semi-circle.

6. Take two sheets of 6-inch square beancurd skin. Put them flat on the big beancurd skin near the flat edge of the semi-circle. Put Nori seaweed and the remaining two sheets of 6-inch square beancurd skin over them. Arrange beef shin slices evenly. Fold the two ends of the big beancurd sheet toward the centre. Arrange shredded ginger pickle on top. Roll it up tightly. Seal the seam with egg white.

7. Wrap the beancurd skin roll in cling film. Prick holes on the cling film evenly. Steam for 30 minutes. Leave it to cool. Refrigerate.

8. Before serving, cut the roll at an angle into 6 pieces. Dribble with thousand island salad dressing and sprinkle with flying fish roe. Put them on the cucumber rings and serve with lettuce leaf.

Ingredients

8 slices marinated beef shin
2 sheets Nori seaweed
40 g young ginger pickle
4 sheets beancurd skin
2 tbsp egg white
8 tbsp flying fish roe
8 tbsp thousand island salad dressing
12 pieces Romaine lettuce
1 large cucumber

Soaking stock

2 cups water
2 tsp sugar
1 tsp chicken bouillon powder
1 tbsp sesame oil
1/2 tsp Maggi's seasoning
1 tsp salt

Tips

· To make marinated beef shin yourself, please refer to the recipe of Spicy soy-marinated beef shin on p.160. Or, you may buy marinated beef shin from stores and slice it yourself.

· After you marinate the beef shin, you have to refrigerate it until firm before slicing. When it's still warm, you can't cut it neatly and it tends to fall apart.

Chicken feet in coconut milk marinade

Ingredients
600 g chicken feet (about 15 feet)
1 smoked coconut
20 g ginger
20 g spring onion

Spice stock
1 kg water
5 g cumin seeds
5 g diced sand ginger
2 cloves star anise
5 g bay leaves
20 g sliced ginger

Marinade
165 ml coconut milk (a small can)
Water from a smoked coconut
50 g table salt
15 g rock sugar
20 g chicken bouillon powder
30 g Shaoxing wine

Dipping sauce
Thai sweet chilli sauce for fried chicken

Method

1. Open the smoked coconut and save the water for the marinade. Set aside the flesh.

2. To make the spiced marinade, boil the spice stock first. Turn to low heat and simmer for 15 minutes. Add marinade ingredients and cook until sugar dissolves. Turn off the heat and leave it to cool. Strain and set aside.

3. Rinse the chicken feet and clip off the nails. In a pot of water, add 20 g of ginger and 20 g of spring onion. Bring to the boil. Put in the chicken feet and bring to the boil again. Turn to low heat and cook for 10 minutes. Turn off the heat and cover the lid. Leave the chicken feet in the water for 35 minutes.

4. Drain the chicken feet. Soak them in ice water until cool and no longer sticky. Chop chicken feet in half. Set aside.

5. Put the chicken feet and smoked coconut flesh into the marinade from step 2. Refrigerate for at least 4 hours. Serve with Thai sweet chilli sauce on the side.

Tips

· Do not cook the chicken feet all the way through on high heat. They may turn too mushy with broken skin that way.

· After cooking the chicken feet, stop the cooking process instantly by plunging them in ice water. The thermal shock crisp up the skin for better mouthfeel.

· Before putting the chicken feet into the marinade, I prefer cutting them in half. Not only do they pick up the flavours more quickly that way, they are also easier for your guests to chew.

廖教賢
Liu Kau-yin, Alvin

廖教賢師傅擁有豐富的飲食業經驗，在本港及外地工作達30多年，曾於香港專業教育學院（酒店、服務及旅遊學系）擔任中菜廚藝導師，現擔任飲食顧問、烹飪中心導師及電視節目廚藝專家。

現為香港餐飲專業技師（國家職業資格）協會常務委員、英國環境衛生協會特許導師，並取得多個協會頒授之專業資格，包括：香港職業訓練局「過往資歷認可」評核員、廣東省勞動和社會保障部高級技師證書（中式烹調師）等，曾獲第五屆全國烹飪技術比賽團體比賽銀獎。

他掌廚之外，積極編撰食譜，曾出版多本烹飪書籍，如《大廚小宴》（榮獲 Gourmand World Cookbook Awards 香港最佳主廚食譜及香港最專業食譜）、《簡材料·滋味餸》、《4 廚具煮好菜》等，並在本港多份報章及雜誌撰文，將入廚心得與讀者分享。

Chef Alvin Liu is an experienced cooking master in the food and beverage industry. Working here and abroad for over 30 years, Chef Liu had been a cookery tutor in Chinese cuisine for Hong Kong Institute of Vocational Education (Hotel, Service and Tourism Studies Discipline). He is now a consultant on food and beverage, a cooking centre tutor, and a cooking expert presenting on television programs.

Apart from being a member of the standing committee of Hong Kong Catering Masters (National Occupational Qualification) Association and an accredited tutor with The Chartered Institute of Environmental Health in the UK, Chef Liu has acquired the professional qualifications awarded by a number of associations, including "Recognition of Prior Learning" assessor of Vocational Training Council, and Senior Technician Certificate (Chinese cooking) from Department of Labor and Social Security of Guangdong Province. He has also won the Overall Silver Medal in group competition of the Fifth National Culinary Competition.

Besides mastering the kitchen, Alvin is enthusiastic in compiling recipes and has published many cookbooks, including *FEAST of Master Chef* (to have a honor to win "The Best Chef Cookbook in Hong Kong" and "The Best Professional Cookbook in Hong Kong" of Gourmand World Cookbook Awards); *Gourmet Cooking with Simple Food*; *Home Favourites with 4 Cookwares*. He also writes for various local newspapers and magazines to share with readers what he has learned from cooking.

陳 永 瀚
Chen Wing-hon, Philip

陳永瀚師傅入行至今，先後在多間食肆任職，包括：西苑酒家、康蘭酒店（蘭苑）、榕樺閣、深井燒鵝海鮮酒家、海天皇宮、全旺海鮮酒家、利苑酒家、陶源酒家、翠園、翠玉軒（米芝蓮一星餐廳）、太平洋會等。近年，他經常參與編撰食譜，著作有《燒味‧傳承滋味》、《巧製燒臘三弄》、《粵式酒樓美食 60》、《百家燒味》等，並擔任餐飲顧問公司及烹飪教室的導師，教授燒味製作的技巧。

從事燒味行業至今二十多年，他曾在不同風格的食肆工作，累積了豐富的烹飪知識，他抱着與時並進、敢於創新的宗旨，以認真的烹飪作風，力求將每個程序更精益求精。

Chef Philip Chen has been working in different restaurants throughout his career, including West Villa, Lan Yuen of Grand Tower Hotel, Yung Wah Court, Sham Tseng Barbecue Goose Restaurant, Citiplaza Harbour Restaurant, Allmax Hotpot & Seafood Restaurant, Lei Garden, Sportful Garden, Jade Garden, Pacific Club and the Michelin 1-star The Square. In recent year, he mainly compiles recipes and has published cookbook including *Authentic Flavours of Barbecue Meat*, *60 Cantonese Delicacies in Hong Kong*, *Celebrity Chefs' Favorite*. He works as an instructor for F&B consultancy companies and cooking classes, specializing in barbecue and marinated meat.

With his 20-plus years of experience in making Chinese barbecue and marinade, Chef Chen has accumulated a profound repertoire of culinary knowledge across different regional cuisines. But he never feels complacent and never stops keeping abreast of the times. True to his innovative instinct, he treats cooking very seriously while fine-tuning every little detail toward perfection.

燒味滷水小吃　Juicy Barbecue Meat, Enjoy!

作者　Author
廖教賢・陳永瀚　Alvin Liu · Philip Chen

策劃/編輯　Project Editor
Karen Kan · Catherine Tam

攝影　Photographer
Imagine Union

美術統籌及設計　Art Direction
Amelia Loh

美術設計　Design
Man Lo

封面設計　Cover Design
Nora Chung

出版者　Publisher
Forms Kitchen
香港鰂魚涌英皇道1065號　Room 1305, Eastern Centre, 1065 King's Road,
東達中心1305室　Quarry Bay, Hong Kong
電話　Tel:　2564 7511
傳真　Fax:　2565 5539
電郵　Email: info@wanlibk.com
網址　Web Site:　http://www.wanlibk.com
　　　　　　http://www.facebook.com/wanlibk

發行者　Distributor
香港聯合書刊物流有限公司　SUP Publishing Logistics (HK) Ltd.
香港新界大埔汀麗路36號　3/F., C&C Building, 36 Ting Lai Road,
中華商務印刷大廈3字樓　Tai Po, N.T., Hong Kong
電話　Tel:　2150 2100
傳真　Fax:　2407 3062
電郵　Email: info@suplogistics.com.hk

承印者　Printer
萬里印刷（香港）有限公司　Prosperous Printing (HK) Company Limited

出版日期　Publishing Date
二〇一四年七月第一次印刷　First print in July 2014
二〇一九年二月第三次印刷　Third print in February 2019